AMERICA'S PREMIER GUNMAKERS

BROWNING

AMERICA'S PREMIER GUNMAKERS

BROWNING

K.D. KIRKLAND

JG PRESS

Published by World Publications Group, Inc.
140 Laurel Street
East Bridgewater, MA 02333
www.wrldpub.net

ISBN 1-57215-101-3
978-1-57215-101-7

Printed and bound in China by
SNP Leefung Printers Limited

Designed by Ruth DeJauregui
Edited by Timothy Jacobs
Captioned by Marie Cahill

Acknowledgements

The author wishes to thank Mr Paul Thomp-
son of Browning NA for his gracious and
invaluable assistance. The author also wishes
to acknowledge the book John M Browning,
American Gunmaker, by John Browning and
Curt Gentry, as a valuable informational source,
parts of which have been adapted for use in the
present book by the gracious permission of
Browning NA. Finally, special thanks go to
Sylvia Kirkland for her help in manuscript
preparation.

Picture Credits

All photos courtesy of Browning NA except:
American Graphic Systems Archives 7, 41,
   43, 97 (all)
Mike Badrocke 70-71, 74-75
The Church of Jesus Christ of Latter Day
   Saints, Visual Resources Library 32 (bot-
   tom)

Colt Firearms 25 (bottom left and right)
Ian V Hogg 46, 47, 72 (top)
© J&J Kidd 42-43, 50-51, 82-83
© KD Kirkland 82 (bottom), 96, 108 (all), 109
Mike Trim 74 (bottom right)
US Air Force 70 (top), 71 (right) 72-73, 74
   (bottom left), 75 (top)
US Army 49

Page 1: John Moses Browning, one of the most
prolific firearms inventors in the history of the
world.
Pages 2-3: These are just two fine examples of
Browning shotguns manufactured by Miroku
Firearms Mfg Co of Japan.
These pages, above and below: A Winchester
Model 1886—which was the first Brown-
ing design ever manufactured by Winches-
ter; a Browning Automatic Rifle, aka BAR.
This particular model is used for big game
hunting.

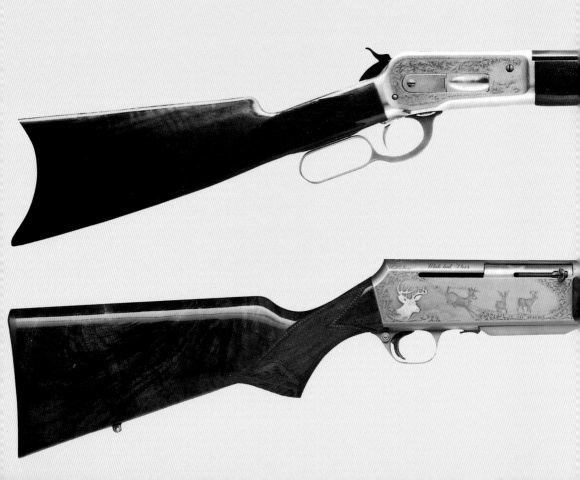

Table of Contents

Jonathan
and His

Above: The Cylinder Repeating Rifle—one of Jonathan Browning's first designs. *Below:* Jonathan Browing's lathe, on which his son John M Browning would learn the art of gunsmithing. *At right:* Abraham Lincoln traded yarns with Jonathan Browning.

Browning Guns

The Father of the Son

Jonathan Browning, destined to be the father of one of the world's great gunmakers, was born on 22 October 1805 at Brushy Fork of Bledsoe Creek in Summer County, Tennessee. He was an industrious, talented fellow, and fixed up his first gun at the tender age of 13, while he was apprenticed to a local blacksmith.

By the time he was 19, Jonathan was considered to be a competent gunsmith, and he forged his own tools for his gunsmithing work. Some of these tools were still in use in the twentieth century, as Jonathan's son John Moses Browning found the old tools also served him well in his shop in Ogden, Utah.

On 9 November 1826, one month after he became 21, Jonathan married Elizabeth Stalcup. Their first child was born in August of the following year, and Jonathan's first gun shop was established at Brushy Fork. He worked there from 1824 to 1834.

While none of them survive, at least some of the guns he made at Brushy Fork saw action in the Seminole Wars of 1836, and in the Civil War (to which many combatants brought their own personal weapons).

Jonathan eventually moved to Quincy, Illinois, then a settlement of 753 people. He invented both his Slide Bar Repeating Rifle and his Cylinder Repeating Rifle in Quincy— descriptions of both of which follow this biography. His gunsmithing business in Quincy flourished, and he entered a period of comfortable prosperity about the time he was elected Justice of the Peace. By the time he was 35, Jonathan had eight children.

Through his lawyerly cousin, Orville H Browning, Jonathan made the acquaintance of a young Illinois lawyer by the name of Abraham Lincoln. Lincoln spent the night in the Jonathan Browning home on at least two different occasions, and the wise, folksy cameraderie between the two men was a Browning family staple for generations.

Also during his middle 30's, Jonathan became interested in religion. The Church of Jesus Christ of Latter-Day Saints,

commonly referred to as the Mormon Church, had its beginnings in New York state. Due to continuous persecution, church leader Joseph Smith led the Mormons to the banks of the Mississippi River and founded a settlement which he named Nauvoo (from the Hebrew word denoting a beautiful place). Within a year they had filled the swamp and turned Nauvoo into a city laid out in large square blocks with over 250 houses. Converts were arriving in large numbers, and in a few short years, Nauvoo became the largest city in Illinois. Sometime in 1840, Jonathan was converted to Mormonism, and shortly thereafter, moved to Nauvoo where he set up a two story brick residence, with the first floor serving as his gun shop.

On 27 June 1844, Joseph Smith and his brother Hyrum were murdered by a mob in Carthage, Illinois. Brigham Young became leader of the Mormons. Figuring that the persecution wouldn't stop, he immediately began to follow through on an old contingency plan to leave what was then the United States and settle the Latter-Day Saints in the Rocky Mountains.

In the midst of this exodus, a message arrived from the President of the United States requesting that the Mormons supply 500 or so volunteers for the war against Mexico which was then raging. Jonathan Browning added himself to the lists of the Mormon Battalion but Brigham Young forbad him to go.

Jonathan's gunmaking skills were far more important to the success of the Mormon migration to the west, than any fighting skills he could offer to the US in the war with Mexico. So it was that Jonathan provided the Utah pioneers with both his Slide Repeating Rifle and his Cylinder Repeating Rifle, both of which operated according to the tenets of simple construction, easy operation and accurate efficiency.

Above: Jonathan Browning, father of John M Browning. *Below:* Jonathan Browning's home and gunsmith shop at Brushy Fork of Bledsoe Creek, Tennessee. *Right:* Jonathan Browning's home in Nauvoo, Illlinois. *Far right:* Map tracing the Mormon Exodus from Nauvoo, Illinois to Salt Lake City, Utah. Jonathan Browning's guns aided the Mormons in their journey westward.

The Brownings and some of their Mormon companions were told to set up a settlement for a while at Kanesville, Iowa, and here they paused in their exodus.

On 24 July 1947, Brigham Young led the advance company of pioneers down into the Salt Lake Valley, and in 1852, Jonathan was permitted to load his wagons and follow the westward trail. Jonathan settled his family in Ogden, Utah. He never again applied himself to the inventing of new guns, as his talents at being a mechanic and engineer were in great demand, due to the constant battle for survival against the elements in that frontier community.

Jonathan and Elizabeth had 11 children by then, and two years after their arrival, he married Elizabeth Clark, who became the mother of John Moses Browning, his brother Matt, and a daughter who died in infancy. A few years later, he married Sarah Emmett, who bore him seven more children, for a grand total of 22! Jonathan Browning died on 21 June 1879 in his 74th year. He had had time enough to become glad in the knowledge that he had sired, in the person of his son John Moses, a greater gunmaker than himself, and had helped to give direction to his talent.

Jonathan Browning's Guns

Slide Bar Repeating Rifle. This is a percussion repeating rifle of approximately .45 caliber, with an ingenious magazine that holds five (or 25) shots. It has a 40.3-inch octagon barrel, and weighs 9.9 pounds.

One of the earliest known American repeating rifles, this firearm was invented by Jonathan Browning sometime between 1834 and 1842 while he was residing in Quincy, Illinois. It was never patented, but was manufactured by its inventor during his residences in Quincy and Nauvoo, Illinois; and Kanesville, Iowa—altogether, from 1842—1852. The total number manufactured is not known.

The Slide Bar Repeater has a number of particularly ingenious features, including its five-shot magazine (a 25-shot variant was available) that takes hand loads. The mechanism centers around a rectangular iron bar with five percussion chambers—for the hand loads—bored in it. This slides through an aperture at the breech and is manually operated.

Since it is an underhammer firearm, the shooter can ease his index finger forward from the trigger to cock the rifle,

thus never having to drop the gun from his shoulder during a firing session. Also, having the percussion chambers located in the bar permits comparatively fast 'mass production' loading, and provides the single-barrel firearm with the capability of firing five (or 25) shots in comparatively rapid succession. A thumb-operated lever, on the right-hand side of the breech, forces the slide against the barrel as each load moves into line with the bore. This creates a gas-tight seal.

The Jonathan Browning Slide Repeating Rifle is not only one of the earliest repeating rifles ever made, it is also one of the simplest—both in its small number of parts and its ease of operation.

Slide Bar Repeating Rifle

Cylinder Repeating Rifle. This is a six shot repeating percussion rifle of approximately .45 caliber, with a six-shot cylindrical magazine. It has a 29.9-inch half octagon, half 16-sided barrel and weighed 12.1 pounds.

The Cylinder Repeating Rifle was invented by Jonathan Browning during his residence in Quincy, sometime between 1834 and 1842. Like the Slide Repeating Rifle, it was never patented. Also, this rifle was not manufactured after Browning arrived in Utah in 1852. The total number manufactured is not known.

The powder and ball are loaded into the cylinder chambers and a percussion cap is placed on each nipple. The rifle is cocked by drawing back the hammer, but there is no mechanism for revolving the cylinder when the hammer is cocked; the cylinder has to be rotated manually after each shot.

The front edge of the cylinder has a tapered cone around each chamber which fits into the breech of the barrel. There is a conical cam in the rear of the receiver, which, when engaged, jams the cylinder tight against the breech of the barrel. This cam traverses horizontally through the receiver immediately to the rear of the cylinder.

John Moses

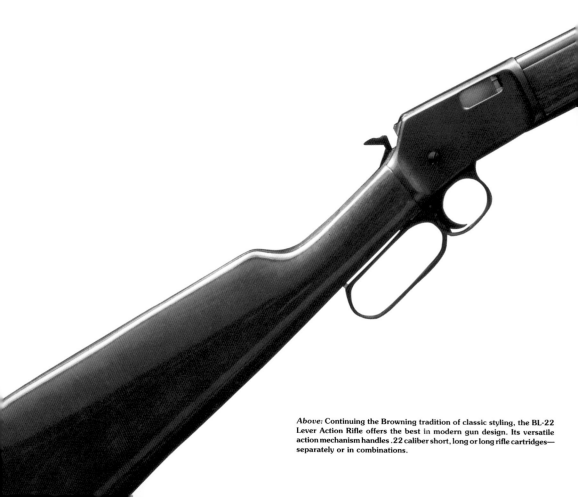

Above: Continuing the Browning tradition of classic styling, the BL-22 Lever Action Rifle offers the best in modern gun design. Its versatile action mechanism handles .22 caliber short, long or long rifle cartridges—separately or in combinations.

Browning

The Man of Inventions

John Moses Browning was born 29 January 1855 in Ogden, Utah. His first toddling steps were toward his dad's gun shop. Family remembrances indicate that he actually started his career at about age six when he dragged a box into the shop to serve as his own work bench and, about a year later, the seven year old was beginning to take himself seriously as a gunsmith. The first gun that John made was composed of the smashed barrel of an old flint lock, a stick of wood, a piece of wire and a scrap of tin.

In 1869, the transcontinental railroad was completed and the Mormons were no longer isolated, but were tangibly connected to both coasts of the country. Just as Utah then stood astride a growing America, so in a few years would John Browning find himself squarely astride new firearms developments. His father had watched the percussion lock gradually replace the flintlock, and now John witnessed the passing of the percussion lock and the introduction of the cartridge firearm.

His father's gun shop was reconditioned in 1873, and it was here in 1878 that John invented the first of his many famous guns—the 1878 Single Shot Rifle, for which John received $8000 from the Winchester Repeating Arms Company.

Thus began a long and profitable relationship between Winchester and Browning. When he was just 27 years old, John invented what would later become the Model 86 Winchester. This was a milestone in firearms design, and it was so efficient that it remained in the Winchester line for over 70 years. In 1887, he invented a lever action repeating shotgun.

John Browning did most of his designing in his own head, and before undertaking the making of a particular model, he had it completely visualized through completion. He never thought that faculty to be unique; he supposed that it was a common attribute. He would, however, use sketches and templates to prove to himself that the designs were indeed sound mechanical concepts.

Although he drew no blueprints, he would make numerous measurements, and would carefully denote them. When the work was going well, John often whistled or sang, and once commented that the sounds of the machines and tools were an orchestra to his ear. Whenever he became stumped over some particular difficulty he fell silent, and when frustration built to a certain point, he simply worked harder as though the tension generated a heat that consumed all obstacles.

Once, he was asked how he managed to create a gun of precise parts and functions with no other instruments of measurement than an inside and outside caliper, a compass (which also served as a scriber), a foot rule graduated down to 64ths and a little spirit level he'd inherited from his father. He smiled and replied, 'Why, I went at the job about the same way I've gone at every other job. I find a good starting place, a fixed point—like the North Star—from which I make exact calculations and then I calculate.'

Once a model was completed, he spent considerable time testing it in the hills behind Ogden and even asked his wife, Rachel, to operate the various prototypes to see if she could

easily work the mechanism. This same inventive process went on all his life, model after model.

As was noted earlier, John's association with the Winchester Repeating Arms Company began with their purchase of his Single Shot in 1883 and lasted 19 years until 1902. During that period John—being freed from the burdens of running an arms-manufacturing shop—was able to devote nearly all of his time to inventing. And invent he did! For nearly 20 years Winchester's new firearms were all Brownings (although never known as such), and 34 of the guns purchased by Winchester were never manufactured; the Winchester line could not have absorbed them, but by purchasing them, Winchester kept them out of the hands of its own competitors.

In effect, Winchester was employing John to work exclusively for them and instead of paying him a fixed salary, he was given his asking price for every gun design he submitted. John's prices were high but at the same time Winchester became the world's leading manufacturer of sporting arms. Browning had more patents for firearms designs than any other American inventor during that period.

Browning invented the Winchester Model 1894 Lever Action Repeating Rifle which has been acclaimed 'the most popular hunting rifle ever built, bar none!' The model 1894 was America's first smokeless-powder sporting rifle. It is interesting to note that all of John's rifle designs for Winchester were designed for black powder cartridges, and yet were later adapted to the much more powerful smokeless cartridges, with only a change of barrel steel. John's philosophy was stated thusly: 'If anything can happen to a gun it probably will sooner or later.' He therefore set his margins of safety far in excess of merely adequate requirements, as if to say, 'make it strong enough, then double it.'

Because of John M Browning's genius, Winchester was the only company in the world with a complete line of sporting arms with not a single model facing a competitor that was even remotely dangerous. By 1900, fully 75 percent of the repeating sporting arms on the American market, both lever and pump action, were of John Browning's invention.

However, Winchester's monopoly was to end in 1902 when its president, TG Bennett, refused to pay Browning royalties for a new invention known as the automatic shotgun. In all of its dealings with Browning the Winchester Company had never paid him a royalty but simply bought his inventions outright. However, an automatic shotgun had no precedent, and Browning figured that it would sell like hotcakes, and righfully wanted in on its sales as a pioneering innovator, while Winchester concluded that it would take too long to perfect, and so refused his offer.

Winchester regretted this mistake, for more than 54 years were to pass before another successful autoloading shotgun was developed. Also, thus was cemented the already extant, and eventually long, Browning association with Fabrique Nationale of Herstal, Belgium—who did accept John Browning's offer.

The automatic shotgun, machine guns, and pistols of John M Browning owed their birth to an event in the fall of 1889. At a shooting match, John's attention was riveted by the clump of sweet clover that stood over to one side of his brother Will's shooting stand: it was forcefully moved by the muzzle blast of Will's gun. John thought, 'Now, why can't we use all that wasted energy to help operate the gun?' When he had rounded up his brothers and they ran back towards the gunshop, everyone knew that John was in an inventing frame of mind.

'Yes, sir!' John exclaimed, from the sheer excitement of the idea that was still expanding in his mind, 'An idea, as

Above and below: Two views of the Winchester Model 1885 automatic shotgun—one of John Browning's greatest inventions. An exquisite modern example of this invention is shown on the *overleaf. Left:* Rachel Teresa Child, shortly before she married John M Browning. *Above:* Matt Browning, John M Browning's brother.

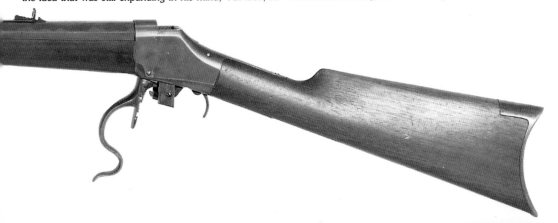

Pappy used to say—biggest one I ever had. Get that damn horse going, Matt.'

'Why, it might even be possible to make a fully automatic gun,' he surmised aloud, 'one that would keep firing as long as you had ammunition.'

As soon as they reached the shop, John took an old .44 caliber Model 73 and wired it to a board, with the rifle lying flat on its side. They nailed the board to the floor. Then they drilled a hole slightly bigger than the old rifle's bore size in a length of two-by-four, and then put it on the floor about a quarter inch from the barrel, with the board's hole lined up with the rifle's bore.

'What happens to that block of wood is what we want to see,' John explained. After a few safety precautions had been arranged they were ready to begin. When they touched the rifle off, the board came to rest only when its momentum had been spent in a ricocheting course of leaps from one obstacle to another. The gas pressure from the rifle proved to be every bit as powerful as John had surmised upon his observation of the sweet clover at the shooting match.

Turning to his brothers he said, 'You know, we may not be more than 10 years away from a pretty good automatic machine gun.' Speaking of that moment later, John's brother Matt said, 'John had Ed and me as excited as a couple of kids at a circus. And then he tells us that we may have a pretty good machine gun, in 10 years.' John was really arguing with himself aloud. He admitted that it did look as though they had stumbled onto something new, the possibility of an automatic gun, operated by the gas that had been wasted since the first shot was fired through a barrel. But, he cautioned, they shouldn't get too excited. 'But,' he concluded, 'it ought to be interesting. Anyhow, Ed, in the morning we'll make a gas-operated gun.''Figure doing it by

noon?' Ed asked with an in-on-the joke wryness. 'Hardly,' John replied. 'About four o'clock I'd guess.' The laughter broke the tension of that momentous day.

On 22 November 1890 John's brother, Matt Browning, sat down and wrote a letter which in time would have tremendous military significance. It was written in long hand on Browning Brothers' stationary and was addressed to Colt's Patent Firearms Manufacturing Company of Hartford, Connecticut. Complete with the author's homespun usages, it read:

Dear Sirs:

We have just completed our new Automatic Machine Gun & thought we would write to you to see if you are interested in that kind of a gun. We have been at work on this gun for some time & have got it in good shape. We made a small one first which shot a 44 W.C.F. charge at the rate of about 16 times per second & weighed about 8#. The one we have just completed shoots the 45 Gov't chge about 6 times per second & with the mount weighs about 40#. It is entirely automatic and can be made as cheaply as a common sporting rifle. If you are interested in this kind of gun we would be pleased to show you what it is & how it works as we are intending to take it down your way before long. Kindly let us hear from you in relation to it at once.

Yours very truly,
Browning Bros

Above right: Two views of Browning's first gas-operated machine gun. *Below right:* Two views of a refined version of the gas-operated machine gun. *Below:* John M Browning with his 37mm aircraft cannon.

Browning's first gas-operated machine gun could fire 600 rounds per minute, while Gatlings of that period had a rate of fire of 1000 rounds a minute but required manual rotation of the mechanism to be fired. By 1895, John's gas-operated machine gun had been developed to such a high degree of efficiency that it was officially adopted by the United States Navy. His first automatic shotgun patent was filed in 1900, and by that year, both Fabrique Nationale and Colt's Patent Firearms Manufacturing Company had Browning Automatic Pistols in production.

All this was accomplished with no curtailment in John's inventing of more conventional rifles and shotguns, as well as miscellaneous inventions such as a machine for loading of machine gun belts invented in 1899 at the request of the Ordinance Board. The Browning Automatic Rifle was officially adopted by the United States in 1917 while the Government .45 Caliber Automatic Pistol had become the official United States military sidearm six years earlier in 1911.

In 1917 the US Government offered John Browning $750,000 for rights to his automatic firearms, and Browning patriotically accepted this sum, knowing that, had the Government paid him standard royalties for the use of these designs, he would have received almost 13 million dollars.

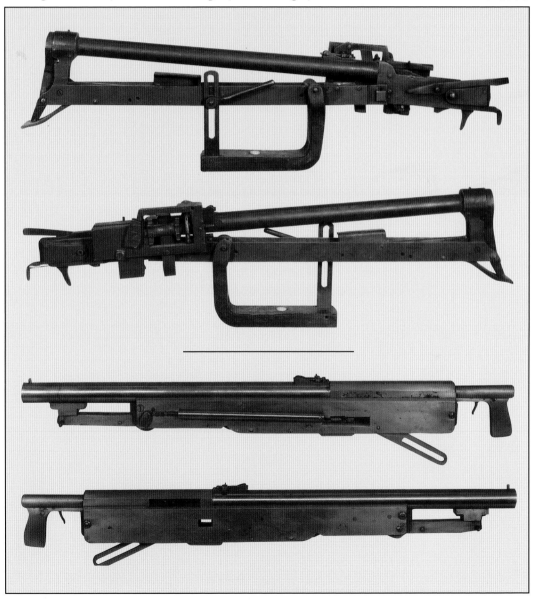

In September 1918, Browning tested his .50 caliber machine gun, which was the 'big brother' to the .30 caliber machine gun that had seen lots of action in World War I. Both of these machine guns saw heavy use—both on the ground, and as aircraft armament—in World War II, and are still in limited use at the present time. In a report from the commanding officer of the US Army Air Force dated November 1943, the .50 caliber Browning machine gun was cited as 'the most outstanding aircraft gun of the Second World War.'

Even Reichsmarshall Herman Goering admitted that 'If the German Air Force had had the Browning .50 caliber, the Battle of Britain would have turned out differently.' To date, no fewer that 66 different known models of the Browning Recoil Machine Gun have been made. The latest aircraft models were stepped up to a cyclic rate of fire of 1300 rounds per minute in the .30 caliber and 1200 per minute in the .50 caliber. Indeed, 'Browning' has come to be a household word. The term 'browning' appears in many French dictionaries as a common noun, uncapitalized, defined as 'one of the pistols designed by the American inventor, John M Browning.'

John M Browning died the day after Thanksgiving 1926 at the Fabrique Nationale plant in Liège, Belgium. The cause of his death was attributed to heart failure. A bronze plaque hangs in the Fabrique Nationale factory in Liège. On it is the likeness of a man, neither young nor old but balding, with a medium-full mustache, an intent serious look in his eyes, and just a hint of amusement in the lines of his mouth. The French on the plaque reads, in English,

Below: **The Browning shop in Ogden, Utah, at about the turn of the century.** *Right:* **The patent drawings for Browning's John M Browning's Single Shot Rifle.**

'To the Memory of John. M Browning 1855—1926. Thirty years previous, he came to this place from Ogden to have his first automatic pistol manufactured, and on the 26th of November, 1926, while he was busily engaged at work, death overtook the greatest firearms inventor the world has ever known.'

In his lifetime John M Browning took out a total of 128 patents covering more than 80 separate and distinct firearms. Upon John Browning's death the presidency of Browning Arms went to John's son, Val, a respected firearms inventor in his own right with 38 patents to his name, and a generous benefactor of colleges and hospitals. Val was even decorated by Belgian royalty. In turn, Val Browning's son, John Val, was president of Browning Arms for 14 years until just before it was acquired by Fabrique Nationale in 1977.

Today, John M Browning's designs remain the staples of the firearms industry, however, the involvement of the Browning family in the company ended with the FN stock buyout. Today, Browning North America, headquartered in Mountain Green, Utah, has divested itself of all product lines not directly related to hunting and fishing.

Firearms manufacturing is split between Asia and Europe with two-thirds of the guns being built by Miroku, a Japanese company that has made only Browning guns for the past 20 years. Fabrique Nationale itself makes the other one third. With total sales approaching $150 million a year, Browning is clearly the market leader in sales of high quality rifles and shotguns as well as home safes.

In January 1988, Browning acquired 37 percent interest in Winchester. Browning Arms has a heritage and a pride in true craftsmanship that continues to satisfy its firearms buffs from generation to generation.

J. M. BROWNING.
Breech-Loading Fire-Arm.

No. 220,271.   Patented Oct. 7, 1879.

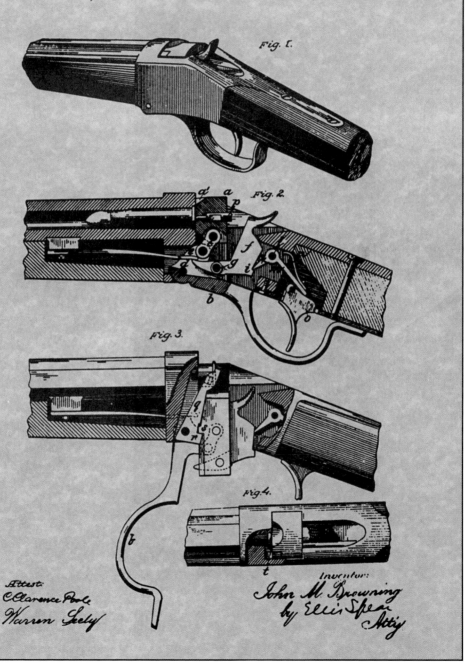

Fig. 1.

Fig. 2.

Fig. 3.

Fig. 4.

Attest:
Clarence Poole
Warren Seely

Inventor:
John M. Browning
by Ellis Spear
Atty

John M
Pistols

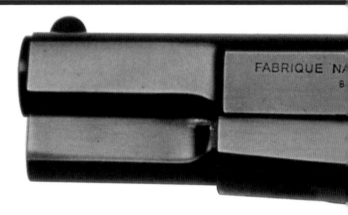

FABRIQUE NA

B

A Selection of John M Browning's Pistols

Semiautomatic .38 Caliber Pistol. This is a gas-operated semiautomatic pistol of .38 caliber, with a pistol-grip magazine, a hammer safety and fixed sights. It has a 5.75-inch barrel and weighs 2.1 pounds. It was the first of John M Browning's many semiautomatic pistol designs and was invented in 1894–95. This pistol is a logical outgrowth of Browning's development of the gas-operation principle, which during the preceding five years he had applied primarily to machine gun models.

A gas vent is located on the top of the barrel a short distance from the muzzle, over which is positioned a piston lever, linked to the breech bolt. As the expanding gases from the detonating cartridge pass through the vent, sufficient pressure is exerted on the lever to force it upward and rearward in an arc. The rearward movement opens the breech bolt, causing it to extract and eject the fired cartridge and cock the hammer. Then, as the lever and breech bolt return to their forward position, a fresh cartridge is fed into the chamber, readying the pistol for the next round.

The patent application for this firearm was filed 14 September 1895, and US Patent Number 580,923 was granted 20 April 1897. On 3 July 1895, the pistol was test-fired by officials of Colt's Patent Firearms Manufacturing Company in Hartford, Connecticut. On 24 July 1896, American manufacturing and sales rights on this pistol (and three others) were assigned to the Colt Firearms Company.

This agreement became the basis for all subsequent agreements between Browning and Colt, even though this particular pistol was never commercially produced. From the day of the agreement, Colt has produced semiautomatic pistols exclusively of the basic Browning design.

.38 Caliber Pistol

Semiautomatic .32 Caliber Pistol. This is a blowback-operated semiautomatic pistol of .32 caliber, with a pistol-grip magazine, a hammer safety and fixed sights. It has a six inch barrel and weighs two pounds. John M Browning invented this pistol in the fall of 1895, and it differs from his first semi-auto design in that the blowback principle, not gas pressure per se, is the basis of its operation. It was test-fired by Colt officials on 14 January 1896, and American

Browning's and Rifles

A weapon renowned for its performance and reliability, the 9mm Hi-Power automatic pistol *above* is one of the many Browning designs manufactured by Fabrique Nationale in Liège, Belgium. The last of John M Browning's pistol designs, it incorporates the improvements and innovations of a quarter century of pistol design.

manufacturing and sales rights were assigned to Colt on 24 July 1896. The patent application was filed 31 October 1896, and US Patent Number 580,926 was granted 20 April 1897. This pistol was never commercially produced.

The expanding pressures within the cartridge case act directly on the breech bolt through the cartridge base, forcing it backward and effecting cartridge ejection and cocking during the rearward movement. A spring located above the barrel provides the energy to slam the bolt shut again, and as the bolt is closing, it catches a fresh cartridge from the pistol-grip magazine, shoving it into the chamber, ready for firing. The slide and breech bolt are integral, and as with any blowback action, must provide sufficient inertia to delay rearward movement until the bullet leaves the barrel and gas pressures have partially diminished.

.32 Caliber Pistol

.38 Caliber Pistol

Semiautomatic .38 Caliber Pistols. These are both semiautomatic pistols of .38 caliber, with positive-lock, recoiling barrels. They each have a pistol-grip magazine and a hammer safety. One of these pistols was designed with fixed sights, and the other with no sights. Both have an 8.9-inch barrel and weigh two pounds. These two pistols were forerunners of the Colt Model 1900. They were invented in 1896 and manufacturing and sales rights were assigned to Colt on 24 July of that same year.

Patent Application Number 67 was filed on 31 October 1896 and US Patent Number 580,924 was granted on 20 April 1897. Patent Application Number 68 was filed on 7 November 1901, and US Patent Number 708,794 was granted on 9 September 1902. Neither of these pistols was commercially produced.

These are the first Browning pistols to employ a positively locked, recoiling barrel. The top of the barrel has transverse ribs and recesses which fit into corresponding ribs in the slide. Each end of the barrel is attached to a link which, when the barrel is in battery position, presses the barrel tightly upward against the slide and thereby interlocks these ribs

and recesses. Thus, in firing, the barrel and slide are locked together with a secure seal at the breach.

In discharging a cartridge, the barrel and slide recoil—locked together until the bullet has left the barrel and gas pressures diminish—at which point the barrel links draw the barrel downward out of the locking recesses, freeing it. The slide alone continues to move rearward, accomplishing the extraction, ejection and cocking functions in the process. The slide is slammed shut by springs, and as it shuts, it chambers a new cartridge from the pistol-grip magazine, readying the firearm for another shot.

Both of these pistols were originally designed to eject from the top, and were later modified to eject from the side.

Semiautomatic .38 Caliber Pistol. This is a short recoil operated semiautomatic pistol of .38 caliber, with a pistol-grip magazine, both hammer and grip safety and fixed sights. It has a 5.9-inch barrel, and weighs two pounds. This pistol was the first pistol to employ Browning's famous grip safety. It was invented by John M Browning in 1896. The patent application was filed on 31 October 1896, and US Patent Number 580,925 was granted on 20 April 1897. The firearm was submitted to Colt in the spring of 1896, and American manufacturing and sales rights were granted the company on 24 July 1896. This pistol was never commercially produced.

This semiautomatic pistol employs a rotating barrel which locks to the slide. It consists of a cylinder-shaped frame which contains a cylinder-shaped slide. The barrel, enclosed by the recoil spring, is wholly contained in the slide. Located near the breech end of the barrel are three pairs of locking lugs which fit into corresponding grooves in the slide and lock the barrel and slide together.

When the gun is fired, the barrel and slide recoil together for a short distance. Upon recoil, two camming studs near

.38 Caliber Pistol

the muzzle cause the barrel to rotate and disengage the locking lugs from their grooves. This allows the slide to separate from the barrel and continue to recoil alone, effecting the ejection of the fired cartridge and the cocking of the hammer. The slide is then slammed shut by its recoil spring, feeding a new new round from the pistol grip magazine into the barrel chamber en route to complete closure.

Model 1900 .38 Caliber Semiautomatic Pistol (Colt). This is a short recoil operated, locked-breech type semiautomatic pistol of .38 ACP caliber having a seven-shot pistol grip magazine, an ingenious rear sight safety and fixed sights. It has a six-inch barrel and weighs 2.2 pounds. From the factory, it was available with plain or checkered walnut grips or checkered rubber grips.

This pistol was the first semiautomatic pistol to be marketed in the United States. The patent application on this firearm was filed 31 October 1896, and US Patent Number 580,924 was granted on 20 April 1897. It was tested by Colt on 29 June 1896, selected for production in

Above: The Model 1900 .32 caliber semiautomatic pistol was the first Browning designed automatic pistol ever maufactured. Produced by Fabrique Nationale, it was immediately popular.

1898 and was placed on the market in February 1900. Its rear sight operated as a safety when pushed down—in which case it blocked the firing pin from the hammer.

Two modified versions appeared in 1902. In the 1902 Sporting Model the safety-sight was replaced with an adjustable rear sight, the hammer was changed from spur to stub round, and as a safety feature a shorter firing pin was used; this last feature soon became a Colt standard. In the 1902 Military Model, capacity was increased to eight rounds, a slide stop was added and the pistol grip was made larger.

Model 1900 .38 Caliber Pistol

Above: **The Colt Pocket Model 1903. Advertisements for the Automatic Colt Pistol** *(left opposite)* **and the New Service Model Ace...a natural for military shooters** *(right opposite).*

**Model 1900
.32 Caliber Pistol**

In 1903 a short-barrel Pocket Model was introduced. The Sporting Models of 1900 and 1902 were discontinued in 1908, the Pocket Model in 1927, and the Military Model in 1928. Total production figures on this pistol are at best approximate. Not only were several numbering systems used for the serial numbers, but large blocks of numbers were set aside for pistols which were—as far as can be determined—never manufactured. Numbering on the military model started at a high figure, then receded, then started back up again. It is estimated that an approximate total of 111,890 of the Sporting, Military and Pocket Models were produced.

Model 1900 .32 Caliber Semiautomatic Pistol (Fabrique Nationale). This is a hammerless, blowback-operated semiautomatic pistol of .32 ACP (7.65mm Browning) caliber, with a seven-shot pistol grip magazine, a thumb safety and fixed sights. It has a four-inch barrel, weighs 1.4 pounds and has factory walnut grips. This was the first Browning-designed semiautomatic pistol to go into production with any firm.

This gun was invented early in 1897. The patent application was filed 28 December 1897, and US Patent Number 621,747 was granted on 21 March 1899. It was shown to Colt officials and a representative of Belgium's Fabrique Nationale shortly after, and on 17 July 1897, a

contract between Browning and Fabrique Nationale was signed, authorizing the firm to manufacture the pistol for all markets outside the United States. Actual production commenced in 1899, thus entitling this design to the above stated claim. The FN .32 Caliber Model 1900, as it was listed by Fabrique Nationale, became immediately popular, and demand was exceptional—100,000 had been produced by August 1904, and 500,000 had been produced by 1909. This model was discontinued in 1910 and was replaced by FN Model 1910, after a total production of 724,450 units.

Model 1911 Government .45 Caliber Automatic Pistol (Colt, Remington and Others). This is a short recoil-operated, locked-breech, semiautomatic pistol which has been manufactured in a wide variety of calibers, including .22, .38, .38 Super, .45 ACP and 9mm. In .45 ACP caliber, this firearm has a seven-shot pistol-grip magazine, comes equipped with manual, grip and magazine safeties and has fixed sights. With a standard barrel of five inches (the original was 3.8 inches), weighs 2.4 pounds, and has a wide variety of grip styles, including diamond-checkered walnut, plain checked walnut and Colt plastic.

Invented in 1905, this is one of John M Browning's most famous designs. The original was the pilot model for what has, for over a half century, been the official United States military sidearm. The patent application for the pistol was

**Military Model
Semiautomatic
.45 Caliber Pistol**

**Model 1911
Government Automatic
.45 Caliber Pistol**

filed on 17 February 1910, and US Patent Number 984,519 was granted on 14 February 1911. A second patent application—covering the details of the mechanical safety—was filed 23 April 1913 and US Patent Number 1,070,582 was granted 19 August 1913.

Colt commenced production in late 1905 with the first models reaching the market in the spring of 1906. The model was considered a commercial success and remained on the market in essentially its original design form until 1911, although some slight modifications had been made in 1909 and 1910.

The US government automatic pistol trials in March 1911 brought in a recommendation for adoption. Upon ratification, adoption became official on 29 March 1911, and the first Military Model was manufactured by Colt on 31 December of that year. A commercial model, identical with the Military Model except for markings, appeared on the market on 9 March 1912.

Slight changes in both of these models have been made over the years. The most important was the traditional Browning improvement of redesigning the pistol's parts so that they served several functions, thereby improving an already efficient design by reducing the total number of parts and simplifying the firearm. These changes were covered by a 1913 Browning patent.

In 1929 Colt brought out the Super .38 Model, which was patterned after the .45 Model 1911, but was chambered for the .38 caliber cartridge. In 1931, .22 caliber was added

to the design roster, with the .22 Ace Pistol. This was followed by two deluxe-grade models in 1933—the .45 National Match Pistol and the .38 Super Match. The Commander, a lightweight model weighing 1.7 pounds, was made available in 1949. This model was available in .45 ACP, .38 Super and 9mm calibers. In 1957, Colt brought out another deluxe .45 caliber target pistol, the Gold Cup National Match. A counterpart in .38 caliber, the .38 National Match, appeared in 1960.

Although these models cover a wide range of grades, all stick closely to the Model 1911 in basic design specifications. One exception is the .22 Service Ace, introduced in 1937. Based on the Model 1911, it contained modifications by another designer which enabled the shooter to get the same recoil or 'kick' with .22-caliber cartridges as with .45 ACPs,

Automatic Colt Pistol

MILITARY MODEL, CALIBRE .45

CAPACITY OF MAGAZINE. 7 Shots.
LENGTH OF BARREL. 5 inches.
FINISH. Full Blued, Checked Walnut Stocks.

WEIGHT.— 32½ ounces.

LENGTH OVER ALL, 8 inches.

Colt "Service Model Ace" Automatic Pistol

CALIBER:
.22 Long Rifle
(Both Regular and High Speed Ammunition)

WITH FLOATING CHAMBER

New Floating Chamber Increases Recoil Approximately Four Times

Specifications

CAPACITY OF MAGAZINE: 10 cartridges.

LENGTH OF BARREL: 5 inches.

LENGTH OVER ALL: 8½ inches.

ACTION: Hand finished.

WEIGHT: 42 ounces.

STOCKS: Checked Walnut.

TRIGGER AND HAMMER SPUR: Checked.

FINISH: Full Blued.

SIGHTS: Ramp front sight, fixed. Rear sight adjustable for both elevation and windage. Both stippled.

ARCHED HOUSING:Checked.

The New Service Model Ace has been designed to provide efficient and economical target practice for military men, and all shooters of the heavy frame Colt Automatic Pistols. It is similar in design to the regular Ace Model . . plus the recently perfected Floating Chamber. By the use of the Floating Chamber the recoil has been increased four times, simulating the recoil found in the .45 caliber Government Model Automatic Pistol. Thus the shooter is trained with an arm that allows him to later change to the heavier caliber pistol, without the additional recoil being noticeable. Because of the much lower cost of .22 caliber ammunition the Service Ace will pay for itself in a very short time.

Special Features

Except for difference in caliber, the new SERVICE MODEL ACE and the Government Model .45 are practically twins. They are so near alike that you can switch from one to the other and hardly notice the difference. However, the Service Ace is provided with hand finished action and a two-way Stevens adjustable rear sight. The front sight is fixed with serrated face.

The Service Ace saves real money and pays for itself in a very short time. It provides accurate, economical target shooting for Service men – members of National Guard, Reserve Officers, and individual shooters of the .45 Caliber Automatic Pistol . . at one-seventh the cost of .45 automatic cartridges.

Above: Designed in 1905 and made by Fabrique National, this was the first .25 caliber semiautomatic. An almost identical model was manufactured by Colt.

Model 1910 & 1922 Pistol

effecting an economy in combat practice ammunition and paying an indirect compliment to the original inventor at the same time.

Manufacture of the two Ace models was discontinued in 1940, but because the stock of parts for these was so large, Ace pistols were sold until 1947. Altogether, 10,935 units of the Acq and 13,800 units of the Service Ace were manufactured. Likewise, the .45 National Match and the .38 Super Match were discontinued in 1940; but they were still numbered with the regular commercial models for years afterward.

Between 1911 and the beginning of World War I, approximately 100,000 of the official Government Model 1911s were produced for the US Armed Forces. In 1917, the Ordinance Department, in an attempt to step up production, contracted with nine companies to produce the pistol. Of these nine, however, only the Remington Arms Company actually entered production before the war ended. For the duration, Colt produced 488,450 and Remington 21,676 units, for a wartime total of 510,126 pistols overall.

During World War II, an estimated 1,800,000 Government Model 1911s were made. Under a licensing agreement similar to that of World War I, Colt and Ithaca each produced about 400,000 units; Remington-Rand, Inc produced about 900,000 units; and the Union Switch and Signal Company produced about about 50,000 units.

After visiting all the factories which, in peace and war, have produced this weapon and examining their records, martial arms authority Lieutenant Colonel RC Kuhn has determined that, just by the end of 1945, total production of the military model was 2,695,212 units—and production has continued 'for lo, these many years since.'

Semiautomatic .25 Caliber Pistol Model Vest Pocket (Fabrique Nationale, Colt, Browning and Others). This is a blowback-operated, hammerless, semiautomatic pistol in .25 ACP caliber, with a six shot pistol-grip magazine, with thumb, grip and (later models) magazine safeties. This pistol has fixed sights, an overall length of 4.5 inches and weighs 13 ounces. Grips on various examples are black rubber with checked field and checked walnut.

This pistol was the first .25 caliber semiautomatic. It was invented by John M Browning in 1905, and was patented in Belgium that same year. This was the only Colt automatic pistol that was truly hammerless—it was discharged by a striker mechanism.

In the United States, patent application on the gun was filed 21 June 1909, and US Patent Number 947,478 was granted on 25 January 1910. The pistol's size partially accounts for (but is by no means proportionate to) its sales. It was first manufactured by FN in 6.35mm in 1905, and approximately 100,000 were sold within the first five years. Colt obtained license to manufacture the pistol in the United

Woodsman
.22 Caliber Pistol

.22 Caliber Long Rifle
Practice Pistol

States and their version, chambered for the .25 caliber ACP cartridge, was introduced in October 1908, with sales numbering 141,000 by 1917. A later Browning Arms Company model, introduced in 1953, also proved highly successful.

The FN and Colt Models are almost identical, except for the thumb safety. The FN safety has a hook which latches into a notch forward of the regular safety notch in the slide and holds the slide to the rear in such a way that the barrel can be easily turned for takedown. The Colt safety has no hook, and has only one notch in the slide which locks the slide when the safety is on. The only other modifications in the Colt model were the addition of a magazine safety in 1917 and a change in sights. The Colt model was discontinued in February 1947, with total production of 420,753 units.

The first FN model was discontinued in 1940, after production of 1,080,408 units. The current Browning and FN Model .25 Caliber followed. In this, the grip safety has been eliminated, and the weight of the pistol has been reduced to 10 ounces in the standard model and 7.8 ounces in the lightweight model; the overall length was reduced to four inches in both. Even before the expiration of Browning's patents, numerous imitations of this gun had appeared on the market, particularly from Spain. Production figures on all models probably number well into the millions.

Woodsman .22 Caliber Semiautomatic Pistol (Colt).
This is a blowback-operated, hammerless (concealed hammer), semiautomatic pistol in .22 long rifle caliber (early models, however, handled only low-velocity .22s), with a 10-shot magazine. This pistol has thumb and automatic (breechblock must be fully closed for firing) safeties, fixed rear/bead or partridge-style front sights, and a (after 1933) 4.5 or (pre-1933) 6.5-inch barrel. With the shorter barrel, the Woodsman weighs in at 10 ounces; with the longer barrel, this figure is 1.75 pounds. Grips for all variants tend to be checkered walnut.

The Woodsman pistol was invented in 1914. The patent application was filed on 30 March 1917 and US Patent Number 1,276,716 was granted on 27 August 1918. Colt began production on 29 March 1915, first calling it the 'Colt .22 Automatic Target Pistol.' Not until 1927 was it given the name by which it is best known, 'the Woodsman.' Somewhat of a rarity for the era in which it first appeared, it utilizes a half-length slide which completely separates from the breech end of the barrel.

'Lesmok or semi-smokeless, lubricated cartridge only' was Colt's caution on ammunition for early models. In approximately 1920, a change in magazine design made possible the use of high-speed ammunition. Many owners converted their models before this. It was relatively easy to make the pistol capable of handling high-velocity cartridges—

by replacing the original mainspring, housing and recoil spring, and by employing an upgraded magazine.

The 4.5-inch barrel Sport Model was introduced in 1933, followed by the Match Target Woodsman in 1938. This was a deluxe model of the Target Woodsman, with an extra-heavy 6.5-inch barrel, larger grips, new trigger, new sights, an added seven ounces of weight. It represented a radical change in design, with a flat-sided barrel and elongated, curved grips.

Manufacture of all three models was discontinued in 1940, but as a result of a plentiful supply of already-manufactured parts, these pistols remained on the market until June 1947, with a total production of 187,423 units.

Two postwar Woodsman models with familiar names were introduced in May 1947—the Target, with a six-inch barrel, and the Sport, with a 4.5 inch barrel. In December of the same year, Match Target Pistols, with 4.5 or six-inch barrels, were introduced. In 1950, Colt brought out an economy model based on the Woodsman; this had a 4.5 inch barrel, and modifications—on the sights, magazine and other aspects of the Woodsman design—distinguished this as an economy firearm. First designated the 'Challenger,' it was renamed the 'Huntsman' in 1955. In 1959, Colt made a few minor changes in the Target Model and redesignated it the 'Targetsman.'

9mm Parabellum Pistol

Semiautomatic Pistol in 9mm Parabellum Caliber.
This is a blowback-operated, hammerless (concealed hammer), semiautomatic pistol in 9mm Parabellum caliber, with a pistol-grip magazine, thumb and hammer safeties, and fixed front/elevation rear, sights. It has a 5.7-inch barrel and weighs 2.3 pounds.

The barrel is adapted to move rearward a short distance in a line parallel with the movement of the breechblock and slide without being locked to the breechblock.

These ornate 9mm Hi-Power pistols were introduced in 1985 to commemorate John Browning's innovative and highly successful automatic pistol design. Engraved on the top of the Classic edition *(right)* is a bald eagle protecting her young from a lynx. On the sides are engraved profiles of an eagle's head and the words 'One of Five Thousand.' The Classic Gold edition *(left)* features similar engraving with inlaid, 18 karat gold highlights and the words 'One of Five Hundred.'

BROWNING ARMS COMPANY MORGAN, UTAH & MONTREAL P.Q.

ONE OF FIVE THOUSAND

Above: Browning's pilot model 9mm pistol. *Right:* John M Browning and his Automatic 22. *Opposite:* A Browning Brothers advertisement in the *Ogden Morning Herald*, touting their wares.

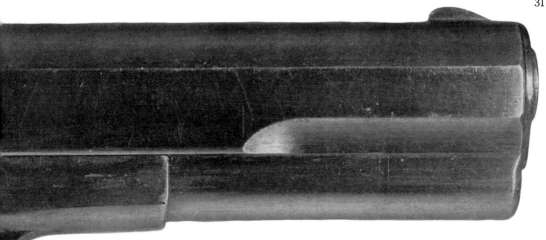

In early 1923, John M Browning was informed that the French Ministry of War was interested in obtaining a semiautomatic pistol of large magazine capacity chambered for the 9mm Parabellum cartridge. Browning completed this model in a few months. This pistol was never patented or commercially produced. It has been seen as a precursor design to Browning's last pistol design, which closely resembled it.

9mm Parabellum Pistol

Semiautomatic Pistol in 9mm Parabellum Caliber (Fabrique Nationale, Browning and Others). This is a short recoil-operated, locked breech, semiautomatic pistol in 9mm Parabellum (9mm Luger) Caliber, with a 13-shot pistol-grip magazine and blade front/fixed rear sights (some military models having graduated-leaf rear sights). This pistol has thumb, hammer and magazine safeties, a 4.6-inch barrel and weighs 2.2 pounds. The grips are checkered wood; military models are also equipped with a quick-takedown shoulder stock which is fastened to the grips so that the pistol can be fired with rifle-like steadiness.

This was John M Browning's last pistol development. The patent application was filed on 28 June 1923, and US Patent Number 1,618,510 was granted on 22 February 1927—three months after John M Browning's death.

This was first produced by Fabrique Nationale in 1935 as the 'Model 1935.' The visible ribs on the breech end of the

barrel are engaged by corresponding grooves in the slide which securely lock the two together upon firing. The breechblock is demountably fixed to the slide, so that the breechblock effectively seals the breech against the force of the fired charge. Recoil continues in this locked position until the bullet leaves the bore. Then a camming action tilts the rear of the barrel downward, freeing the slide and the breechblock to continue rearward, effecting the necessary ejection and cocking operations. A spring slams the action shut anew, and as this operation proceeds, a fresh cartridge is shoved from the magazine into the firing chamber by lugs on the moving mechanism.

The Model 1935 was adopted as the official sidearm of the Belgian Army and other European and colonial troops. Over 200,000 were manufactured in Canada for the Chinese Army during World War II. This pistol was, for years, the standard military sidearm of many of the NATO countries, and the Browning Company sold it as a personal and sporting arm as well.

A Selection of John M Browning's Rifles

Single Shot Rifle (Browning Brothers and Winchester). This is a single shot falling block rifle which has been adapted to a very wide variety of calibers, ranging from .22 short to .50-90 Sharps. It has both hammer and action safeties (half-cock notch and full-close firing only), windage and elevation-adjustable rear/blade front sights, a variety of barrel lengths from 15 to 30 inches (depending upon model) and, likewise, weight varying from 4.5 to 13 pounds. Since this rifle was made in a wide number of variants, stock styles varied widely also, from a standard rifle type stock to the ornate and heavily styled Schuetzen target model, to the shotgun style featured on that particular variant! Barrels likewise varied—in length, as is noted above, and in configurations from round to octagonal to half-octagonal.

This was John M Browning's first firearm, invented in 1878 when he was 23 years old. The patent was filed 12 May 1879, and US Patent Number 220,271 was granted on 7 October 1879. Production by the Browning Brothers, in Ogden, Utah Territory, began about 1880 and continued until 1883, with a total of approximately 600 rifles manufactured. Manufacturing and sales rights were sold to the Winchester Repeating Arms Company in 1883 and the rifle appeared in 1885 as the Winchester Single Shot Model 1885.

The hammer dropped down with the breechblock when the trigger guard was levered open, and cocking was accomplished by the closing movement. Of course, it could also be cocked by hand.

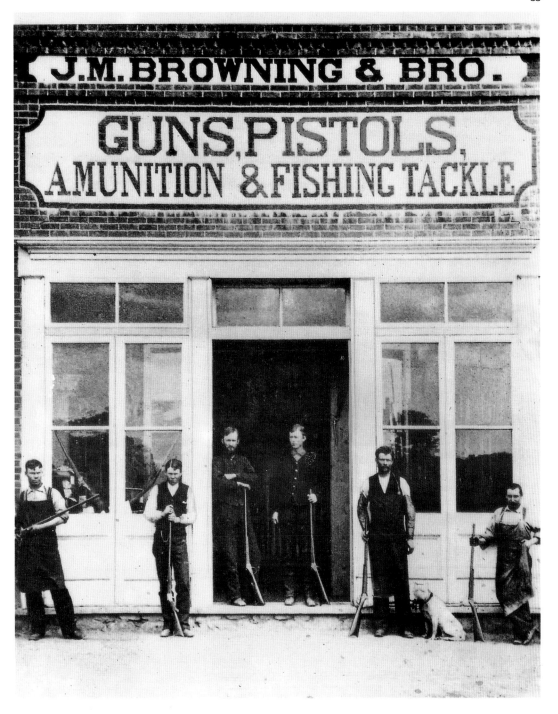

Above left: The inventor at age 18. *Opposite:* The Browning shop and factory as it appears today, and *(above)* as it appeared in 1882. Shown here from left to right are Sam Browning, George Browning, John M Browning, Matthew S Browning, Ed Browning and Frank Rushton.

Overleaf: The Browning tradition of innovation and craftsmanship continues today—*(from the top down)* an A-Bolt 22 Bolt Action, a BL-22 Lever Action Grade II, a BL-22 Lever Action Grade I and a 22 Semi-Auto.

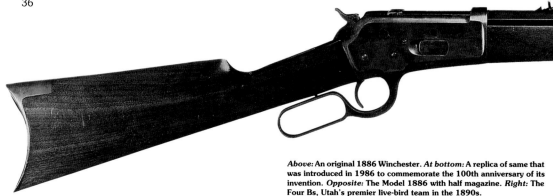

Above: An original 1886 Winchester. *At bottom:* A replica of same that was introduced in 1986 to commemorate the 100th anniversary of its invention. *Opposite:* The Model 1886 with half magazine. *Right:* The Four Bs, Utah's premier live-bird team in the 1890s.

The Single Shot has been adapted to over 33 different calibers, including both rim- and centerfire cartridges—more than any other single shot or repeating rifle known. It was the first Winchester rifle capable of handling the more powerful metallic cartridges of the period.

Through the years the Single Shot was produced in a variety of models. The light carbine (called the 'Baby Carbine') appeared in 1898. The takedown model was introduced in 1910. A special military target version was introduced in 1905; in 1914 it was revamped as the Winder Musket, named in honor of Colonel CB Winder, and was used for training expeditions in World War I. In 1914, the Single Shot was also made into a shotgun, chambered for the three-inch 20-gauge shell.

This firearm was discontinued in all models in 1920. Total production of all models was approximately 140,325, which includes the 600-unit production by the Browning Brothers.

Tubular Magazine Repeating Rifle. This was a bolt action, tubular magazine firearm of indeterminate caliber. The second arm invented by John M Browning was a tubular magazine repeating rifle. Patent was filed 29 March 1882, and US Patent No. 261,667 was granted 25 July 1882. This gun was never manufactured, and no known models survive.

This was probably John M Browning's first tubular magazine design. From the patent application, its bolt action was to have a rotating sleeve with locking shoulders in the receiver. The then-features of this gun were the arrangement of the tubular magazine under the barrel; the receiver, open at the top; and the carrier for elevating cartridges from the magazine to the chamber.

Cartridges are loaded into the magazine through the top of the receiver when the bolt is open. A system of grooves on each side of the inner walls of the receiver guide the cartridges as they are manually forced one by one into the rear opening of the magazine. A spring system in the magazine forces a fresh cartridge onto the spring-loaded carrier at the bottom of the chamber, where, when the bolt is opened, the cartridge is snagged by a lug on the bolt and is thereby slid into the breech when the bolt was closed.

The striker-type firing pin has a finger hook for manual cocking. A spent cartridge is ejected by opening the bolt, which action also allows a fresh cartridge from the magazine to come into position for bolt-loading into the breech.

Lever Action Repeating Rifle. This is a lever action rifle in .45 caliber, with a tubular magazine, an automatic safety and a 26.8-inch octagonal barrel. This rifle weighs 8.8 pounds. It was invented in 1882. Patent was filed 13 September 1882, and US Patent Number 282,839 was granted 7 August 1883. It was never manufactured.

It is a unique firearm in that it operates quite differently from most lever actions. Of very simple construction, the breechblock and finger lever are essentially of one piece. The breech piece is hung to the receiver by the extractor in such a manner that the operating movement causes the rear end of the breech to drop, thereby unlocking the piece, while the forward end, in the receiver, is guided longitudinally by the extractor.

The swinging movement of the combination breechblock and finger piece extracts and ejects the fired shell, and pivots a one-piece carrier upward from the magazine, with a fresh round ready to load when the action is closed. Cocking the hammer is not done automatically, but instead must be done

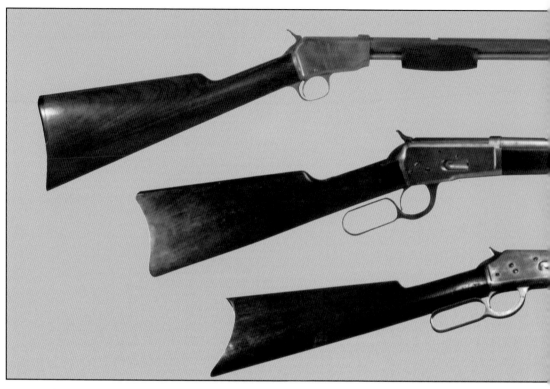

manually for each round, thus constituting an automatic safety.

When in closed position, the rear end of the breech rests squarely against the recoil-bearing surface in the receiver, holding the front end of the breechblock against the barrel. The purpose of this construction is to provide a lever action rifle of extreme simplicity, with a minimum of parts.

Model 1886 Lever Action Repeating Rifle. This is a lever action repeating rifle in a large variety of calibers— including .45-70 US Government, .40-82 WCF, .45-90 WCF, .40-65 WCF, .38-56 WCF, .50-110, .40-70 WCF, .38-70 WCF, .50-100-450 and .33 WCF—with a choice of either full-length or half-length tubular magazine of varying capacities, depending on the ammunition used. Safeties on this firearm are manual and mechanical, and barrels include 26-inch round, octagon or half octagon, 22-inch round and special barrels and magazines that were finished to owner specifications until 1908. The weight of this arm varies widely depending on specifications and caliber, and stock types range from the sporter with straight or pistol grip, to extra-lightweight, to shotgun, carbine and musket-type stocks.

Invented in 1882–83, this was the first Browning-designed repeating rifle ever manufactured. It was also the first repeating rifle from any source to successfully employ sliding vertical locking lugs—which effectively seal the breech and barrel of the gun. By this same distinction, it was the forerunner of all later Browning lever action rifles. It has been said that practically every improvement since made in lever action rifle design was derived from this design.

Patent was filed 26 May 1884, and US Patent Number 306,577 was on granted 14 October 1884. Purchased by Winchester in October 1884, it appeared on the market in 1886 as the legendary Winchester Model 1886. In 1894, the Model 1886 was converted to a takedown model, and in 1936, it was slightly modified to handle the .348 Winchester cartridge, and became the Model 71.

The Model 1886 was discontinued in 1935 with 159,994 produced, and the Model 71 was discontinued in 1957 with 43,267 produced. The total production for this rifle was 203,261 units, and its production life had spanned 71 years.

Model 1890 .22 Caliber Pump Action Repeating Rifle (Winchester). This is pump action repeating rifle of .22 caliber (short, long, long rifle and WRF) with a tubular magazine of varying capacity (from 15 shorts to 11 long rifles), manual and mechanical safeties and a wide variety of sights. It has a 24-inch octagonal barrel and weighs, depending on stock styling, 5.8 to 6.0 pounds. Available stocks include a standard rifle type, with curved steel butt plate and straight grip, and optional pistol-grip stocks.

The patent application on this gun was filed 13 December 1887, and US Patent Number 385,238 was granted on 26 June 1888. The firearm was produced, and appeared in 1890 as the Winchester .22 Caliber Repeating Rifle Model 1890, and was the first repeating pump action gun manufactured by Winchester. It has been called 'the most popular .22 caliber pump action rifle ever made.'

What distinguished this rifle over previous .22 caliber repeaters was the cartridge carrier-feed mechanism. Pre-

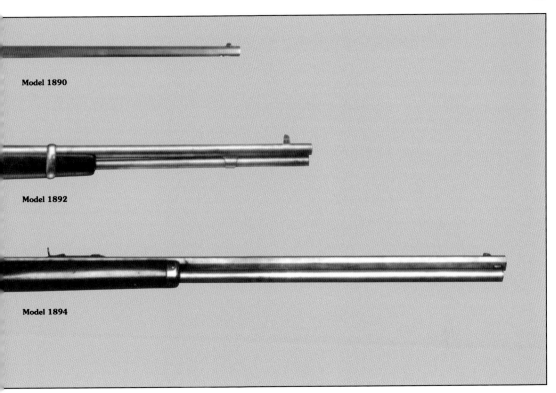

Model 1890

Model 1892

Model 1894

viously, no positive method of handling the .22 caliber short cartridge had been developed. This small cartridge was prone to jam most repeater rifle mechanisms—it was so short that, often, two cartridges would find their way to where only one of them should have been.

Browning's design features a fingerlike cartridge stop on the front of the carrier, which assures that only one .22 short cartridge at a time can fit onto the carrier. When a spent cartridge is ejected from the chamber, the carrier presents the fresh cartridge, ready to be loaded by the mechanism into the chamber.

First manufactured with its barrel permanently mounted to its frame, the Model 1890 was converted to takedown in 1893. The Model '06, introduced in 1906, represents a modification which enabled one rifle to accept all .22-caliber cartridges except the slightly irregular .22 WRF (the Model 1890 versions were not interchangeable). Also, with these changes, a 20-inch round barrel replaced the 24-inch octagonal barrel. In 1932, the Models 90 and '06 were renamed the 'Model 62,' with the introduction on both of a barrel with slightly different specifications, and new sights.

The Models 90 and '06 were discontinued in 1932 with 849,000 and 848,000 produced respectively. The Model 62 was discontinued in 1958 with 409,475 produced. The production total for all variants of this rifle are 2,106,475 units, with a production life of 69 years.

Model 1892 Lever Action Repeating Rifle (Winchester). The Model 1892 is a lever action repeating rifle of either .44-40, .38-40, .32-20 or .25-20 caliber, with a tubular magazine having capacities ranging from five to 17 rounds, both mechanical and manual safeties, and adjustable sights. With barrel lengths ranging from an absolute minimum of 14 inches (special model) to an absolute maximum of 36 inches (special model), the weight of this firearm varied from 5.5 to eight pounds. The Model 1892 was available with a wide variety of stocks.

The 1892 was originally designed with its barrel permanently mounted to its frame, but in 1893 it became available in a takedown model. A modified version with a decreased magazine capacity was introduced in 1924 as the Model 53. Its successor, the Model 65, was introduced in 1933.

The Model 53 was discontinued in 1932 with a production total of 24,916 units; the Model 65 was discontinued in 1947, with a total of 5704; and the Model 92, though not produced except in the carbine model for several years after the introduction of the Model 53, was not officially discontinued until 1941—at which time 1,004,067 had been manufactured. The manufacturing total for all variants combined stands at 1,034,687 units.

Model 1894 Lever Action Repeating Rifle (Winchester). This is a lever action rifle in .32-40 and .38-55 (both black powder), and .25-35, .30-30 and .32 Special (all three, smokeless powder) calibers, with a tubular magazine of various capacities ranging from three to eight cartridges. This design has both manual and mechanical safeties, and adjustable sights. Barrel lengths and styles include 20-inch round, 22-inch round and 26-inch round, octagonal or half octagonal, and weights range from 5.8 to 6.3 pounds for carbine models, from seven to 7.8 pounds for other variants. Stock styles are many and various.

Often called 'the most famous sporting rifle ever produced,' the Model 94 is perhaps best-known as the 'Winchester .30-30.' Patent was filed 19 January 1894, and US Patent Number 524,702 was granted on 21 August 1894, and the Model 1894 was first manufactured by Winchester in 1894—hence, its company monicker. It was revolutionary in that it was the first repeating hunting rifle to handle the smokeless-powder cartridges.

First manufactured with a permanently attached barrel, it became available in a takedown model in 1895. Modified versions include the Model 55, introduced in 1924 (principal differences being a shorter barrel, a redesigned stock and a switch to half-length magazine), and the Model 64, introduced in 1933 (the main difference being in the type of steel used). The Model 55 was discontinued in 1932, and production for the Model 64 halted in 1957.

Still in production as the Model 94 carbine, this rifle has outsold any other manufactured by the Winchester Repeating Arms Company.

Model 1895 Lever Action Repeating Rifle (Winchester).

This is a lever action repeating rifle in a variety of calibers, including .30 US Army, .30-40 Krag, .38-72, .40-72 Winchester, .303 British, .35 Winchester, .405 Winchester, .30 Government 1903, .30 Government 1906 and 7.62mm Russian. This firearm has a four to six round box magazine, adjustable sights and both manual and mechanical safeties. Barrel lengths and styles vary from 22 to 36 inches, and round, octagonal or half-octagonal. Weight and stock style varies according to variant specifications.

This was the first non-detachable box-type magazine rifle designed to handle jacketed sharp-nosed bullets. The patent application for this rifle was filed on 19 November 1894, and US Patent Number 549,345 was granted on 5 November 1895. The firearm was first manufactured by Winchester in 1896, and is known as the Winchester Model 95.

Four slightly different versions of the musket appeared between 1895 and 1908—they are as follows. The .30 Army Model/1895 US Army Pattern was adopted by the US Army in 1895. The same year, some 10,000 of these were chambered for the .30-40 Krag cartridge and were purchased by the US Army for use in the Spanish-American War. Prior to America's entry into World War I, 293,816 of these guns were chambered for the 7.62mm Russian cartridge, and were sold to Russia. The fourth of these particular variants was a takedown version that appeared in 1910.

All models were discontinued in 1931, with a total production of 425,881.

Model 1900 Bolt Action Single Shot .22 Caliber Rifle (Winchester).

This was a bolt action, single shot rifle of .22 (long and short, interchangeable) caliber with a manual safety, adjustable sights and an 18-inch round barrel. This rifle weighs 2.8 pounds.

The patent application on this gun was filed 17 February 1899, and US Patent Number 632,094 was granted 29 August 1899. It was first listed in the Winchester 1899 catalogue as the Winchester Model 1900 Single Shot Rifle. Designed as a low price, single shot 'plinking' rifle, it was of especially simple construction and has been widely copied.

The Model 1900 was discontinued in 1902; the Model 1902, announced the same year, had a modified trigger guard shape, a short trigger pull, a steel butt plate, a rear peep sight and a slightly heavier barrel. In July of 1904, another slightly modified version appeared. This was the Model 1904, which had a longer, heavier barrel and a modified stock. Another interesting variant is the Model 99 Thumb Trigger Rifle, which also appeared in 1904. This rifle has no conventional trigger; just behind the cocking piece on the bolt is a button called the 'thumb trigger.' When ready to discharge a round, the marksman merely presses down on this button with the thumb to release the firing pin.

In 1928 and the years following, Winchester brought out other variations—the Models 58, 59, 60 and 68. In 1920, a shotgun version, similar to the Model 1902, was announced—this being the Winchester Model 36 Single Shot Shotgun. It was the only American-made shotgun chambered for 9mm paper shells, and was discontinued in

Above: An original Model 95 Lever Action Takedown Rifle. *Below:* An original Browning Single Shot, introduced as the Winchester Model 1900. *Right:* An advertisement for the .50-100-450 cartridge, designed especially for the Winchester Model 1886.

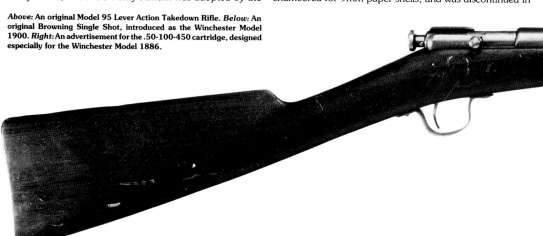

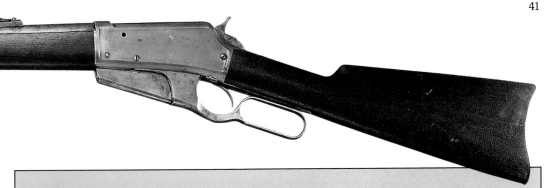

WINCHESTER = Model
1886

.50-100-450.

—A New Cartridge.—

.50 Caliber,
100 Grains Powder.

450 Grain
Solid Bullet.

LIST PRICE, $48 PER 1000.

To meet the demands of our friends for a .50 caliber carrying a heavy bullet, we are now prepared to furnish the above. The bullet has a penetration of about 16 pine boards ⅞ inch thick, and a trajectory of about 12 inches at 200 yards. This cartridge cannot be fired with good results out of the .50-110-300 rifle, but requires a barrel especially rifled for it.

Winchester Repeating Arms Company,
NEW HAVEN, CONN.

Send for our 112 page catalogue—free.

These two Browning guns were purchased by Mr OE Brownsey, a good friend of Val Browning. *Above:* An over and under with solid field rib, ivory sight and single selective trigger. *At top:* This automatic has a semi-beaver tail fore end of burled walnut with fine checkering, a solid field rib with ivory sight and a walnut butt stock with matching checker pattern. *At left* holding the automatic is Everett L Brownsey, OE Brownsey's son. Everett Brownsey served as mayor of Tombstone, Arizona. *Right:* From an earlier era, a group of Dodge City gunfighters who relied on their Browning designed Winchesters to enforce the law in Tombstone in the days of the Wild West.

1927. The last variant, the 68, was introduced in 1934 and was discontinued in 1946. The overall production total, including every variant, stands at 1,458,666.

Semiautomatic High Power Rifle (Remington and Fabrique Nationale).

This is a recoil-actuated, autoloading rifle in .25, .30, .32 or .35 Remington caliber, with a five-round nondetachable clip magazine, manual safety and adjustable sights. It has a 22-inch barrel and weighs 7.8 pounds. The stock for most examples tends to be of the rifle type with a straight or semi-pistol grip, having a shotgun-style rubber butt plate.

This was the first successful autoloading, high-power rifle in the United States. The basis of its autoloading, semiautomatic mechanism is a rotating bolt head having double lug locks in the barrel extension. The barrel itself recoils inside of a barrel jacket while it is still locked to the bolt. A stop-open latch holds the breechblock open after the last shell is fired.

The patent application for this gun was filed on 6 June 1900, and US Patent Number 659,786 was granted on 16 October 1900. US manufacturing and sales rights were granted to the Remington Arms Company, and the rifle first

At top: **Browning's original for the Semiautomatic High Power Rifle.** *Below, from the top down:* **Two of the best known American military arms, the BAR and the Government .45 Caliber Automatic Pistol.** *Opposite:* **John and Matt Browning on an elk hunt.**

appeared in 1906 as the Remington Model 8 Autoloading Center Fire Rifle. Fabrique Nationale introduced the gun in Belgium in 1910 as the FN Caliber .35 Automatic Rifle.

The Remington Model 8 was discontinued in 1936, and was replaced by the Model 81 Woodmaster the same year. Modifications introduced by the Woodmaster included an improved stock and a slight weight increase to eight pounds. The Model 81 was discontinued in 1950. The FN model differed from the Remington models in that it had a solid-matted rib barrel, a bead front sight and two-position-folding rear sight (the Models 8 and 81 had adjustable open rear sights), a checked forearm and buttstock and a weight of 8.3 pounds. The FN model was discontinued in 1931, with a total production of 4913 units. As a matter of company policy, production figures are not available on Remington arms.

The Browning Automatic Rifle (Colt, Winchester, Marlin-Rockwell, Fabrique Nationale and others).

This is an air-cooled, gas-actuated, automatic rifle in a variety of calibers, including .30-06, 6.5mm, 7mm, 7.62mm and 7.9mm, and having a 20 to 40-round box magazine and complex safety arrangements which are described below. It can be fired at a maximum full-automatic rate of 480 rounds per minute, emptying a 20-round magazine in 2.5 seconds. The World War II-era variant, the M1918A2, has a high rate of 550 rpm.

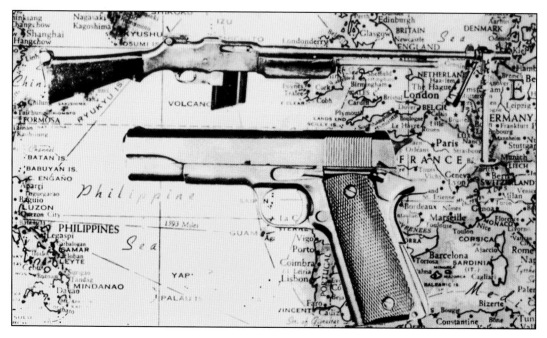

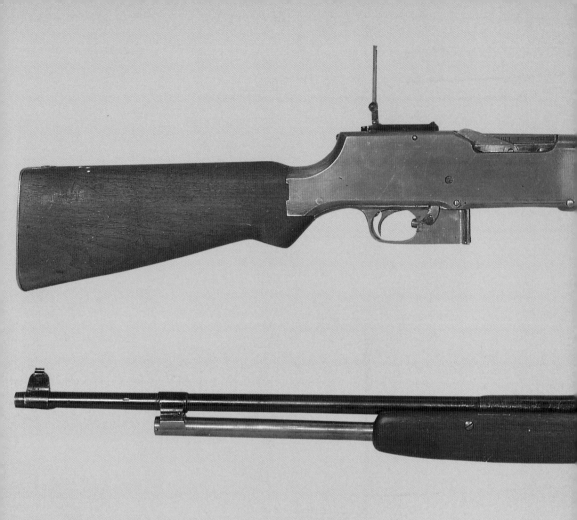

Model 1918A1

At top: Two views of John M Browning's automatic rifle, made in his shop in Ogden, Utah. A forerunner to the US Automatic Rifle Model 1918, this rifle has uncontrolled top ejection. Another variant of the BAR, the Model 1918A1 *(far left)* has a hinged butt plate and bipod attached just ahead of the fore end stock. *Left:* The Browning Automatic Rifle in action.

Best known as the 'BAR,' it is also known as the Browning Light Machine Rifle Model 1917, the Light Browning, the Colt Automatic Machine Rifle and the Fusil Mitrailleur Browning. This rifle has a 24-inch barrel and weighs approximately 20 pounds with a full magazine. Sights are adjustable, and the stock is a heavy rifle type arrangement.

The safety for this automatic weapon is a fire-control change lever. When the change lever is in its forward position, marked with the letter 'F,' the rifle will shoot one shot with each pull of the trigger. In vertical position, marked with the letter 'A,' the rifle will fire at full automatic. In the rearward position, marked by the letter 'S,' for 'safe,' the weapon will not fire.

On some models, when the 'F' lever is all the way forward, the rate of fire is reduced in such a way that single shots may be effected by jerking and releasing the trigger. On the M1918A2, there is no provision for semiautomatic fire, but this last-mentioned method for firing single shots can be used. The M1918A2 also differs from earlier models in that it is slightly heavier and is equipped with a flash hider and bipod.

The patent application for the first of these designs was filed on 1 August 1917, and US Patent Number 1,293,022 was granted on 4 February 1919. The BAR was officially adopted by the United States Government in 1917 and first saw combat use in July 1918. It is designed for takedown in

Below: **John and Rachel Browning at home, a few months before his final trip to Belgium.** *Opposite:* **American troops in Europe during WWII. US forces used the BAR during both World War I and II.**

combat conditions, and its 70 pieces can be completely disassembled and reassembled in 55 seconds.

The BAR has a bolt lock which is pivoted to the rear of the bolt and which rises in and out of locking engagement with a shoulder on the receiver. The rear of the bolt lock is attached to the slide by a link, which is free to reciprocate backward and forward. Attached to the forward part of the barrel is a gas piston which derives its energy from a gas port drilled through the barrel wall. In operation, the piston sends a slide to the rear, and in turn, the slide, through its link connection with the bolt lock, pivots the bolt lock downward—out of locking contact with the receiver.

During World War I, approximately 52,000 BARs were manufactured by Colt, Winchester and Marlin-Rockwell. After World War I, production rights reverted to Colt, and also, by arrangement with John M Browning, Fabrique Nationale began European production in 1920. The FN model is called the Fusil Mitrailleur Leger. Large quantities of this weapon were manufactured for various European countries over the years. It has been widely copied; many nations adopted the BAR or a have had a similar gun in reserve.

In 1922, the US Army brought out its Cavalry Model. In 1933, Colt produced the 'Colt Monitor' for police and bank guard use. In November 1939, there were approximately 87,000 in our war reserve, and approximately 177,000 were produced in this country during World War II. The FN version differed from the Colt model chiefly in having a quicker takedown mechanism—which allowed the barrel to be removed easily for replacement.

John M

A Selection of John M Browning's Shotguns

Model **1887 Lever Action Repeating Shotgun (Winchester).** It is a lever action repeating shotgun in 10 or 12 gauge, with a four-shot tubular magazine (plus one in the chamber), and groove and bead sights. Examples of this shotgun type have 20, 30 or 32-inch barrels, with cylinder bore, riot, standard or full choke. The 10 gauge variants weigh approximately nine pounds, and the 12 gauge variants weigh approximately eight pounds. Stocks for these guns are standard plain with pistol grip and hard rubber butt plate.

This was the first lever action repeating shotgun made in the United States. It has been called 'the first really successful repeating shotgun.' The patent application on this gun was filed on 15 June 1885 and US Patent Number 336,287 was granted on 16 February 1886. Manufacturing and sales rights were sold to the Winchester Repeating Arms Company in 1886, and the manufactured gun appeared in June 1887 as the Winchester Model 1887 Shotgun. Riot gun variants in 10 and 12 gauge were brought out in 1898.

The model 1887 was discontinued in 1899. Redesigned to handle the new smokeless powder loads, it reappeared in 1901—in 10 gauge only—as the Model 1901. This model was discontinued in 1920. Some 64,855 units of the Model 1887 were produced, as were 13,500 units of the Model 1901, for a grand total of 78,355 units overall.

Model 1893 Pump Action Repeating Shotgun (Winchester). This is a pump action shotgun of 12 gauge, with a five-shot tubular magazine and bead and groove sights. Barrel lengths for this gun are 30 and 32 inches, with standard full choke and optional cylinder-bore and modified

Browning's Shotguns

John M Browning's revolutionary automatic shotgun made firearms history even before it was manufactured by precipitating a break between Browning and Winchester. Invented in 1900, the design has remained popular to this day. The sleek Automatic-5 *above* is a current production model. Browning *(at left)* holds an early model of this legendary shotgun.

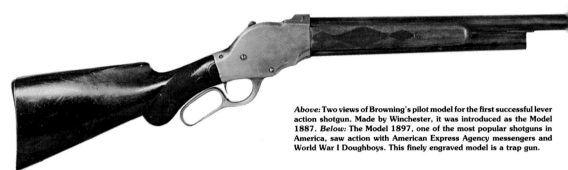

Above: Two views of Browning's pilot model for the first successful lever action shotgun. Made by Winchester, it was introduced as the Model 1887. *Below:* The Model 1897, one of the most popular shotguns in America, saw action with American Express Agency messengers and World War I Doughboys. This finely engraved model is a trap gun.

variants available. The weight for all models is approximately 7.8 pounds, and stocks tend to be plain, with pistol grip and hard rubber butt plate, with options having been available.

This was the first shotgun with a sliding-forearm pump action manufactured by Winchester. The patent application for this gun was filed on 30 June 1890, and US Patent Number 441,390 was granted on 25 November 1890. Manufacturing and sales rights were sold to the Winchester Repeating Arms Company in 1890; the gun was announced in April 1894 as the Winchester Model 1893.

The Model 1893 was discontinued in 1897 when the Model 1897—a much-modified takedown version of the Model 1893—was introduced. Total production for the Model 1893 was 34,050 units.

Model 1897 Pump Action Repeating Shotgun. This is a pump action shotgun in 12 and 16 gauge with a five-shot, tubular magazine, and bead and groove sights. Variant examples had barrel lengths of 20, 22, 26, 28, 30 and 32 inches (plus other variants), with cylinder bore, modified, intermediate, full or Winchester skeet chokes. The weights of the various examples of the Model 1897 are from 7.1 to 7.9 pounds, and standard stocks are plain, with pistol grip and hard rubber butt plate, with optional checkering and other specialties having been available.

This was one of the most popular shotguns in America. Introduced in November 1897, the Model 1897 is a modified version of the Model 1893, with a stronger frame, side ejection, and takedown capability. In addition to its legendary record as a sporting arm, the Model 1897 also saw other usage. It was widely used as a law-enforcement riot gun, and the American Express Agency armed its messengers with

Model 1897s for a time. During World War I, this weapon was used by American forces as a trench gun with much effectiveness.

Back in 1897, the Standard, Trap, Pigeon and Brush Gun variants were introduced, and in 1898, the Riot Gun variant was brought out. This was followed by the Tournament Model in 1910, and the Trench Gun—for Armed Forces use during World War I, and for the public in 1920. The Trap Gun was succeeded by the Special Trap Model in 1931, and the Brush, Riot, Trench and Pigeon models were discontinued in 1931, 1935, 1935 and 1939, respectively. The Tournament Model was succeeded in 1931 by the Standard Trap Model, which itself was discontinued in 1939.

The Standard Model 1897 was discontinued in 1957. After all was said and done, the overall production total for all Model 1897 variants was 1,240,700.

Automatic Shotgun (Fabrique Nationale, Browning, Remington and others). This is a long recoil-operated automatic shotgun. The following specifications refer to the most representative group of variants, the Browning Automatic-5 family of shotguns. These come in gauges 12, 16, 20 and three-inch Magnum 12, with a five-shot magazine, except in three-inch Magnum, which has a five-shot capacity with Folding Crimp three-inch shells and a four-shot capacity with Rolled Crimp three-inch shells. Barrel lengths for the Auto-5 run from 26 to 32 inches, depending on type specifications and option package, and include all possible choke borings, including interchangeable, screw-in, choke systems. Weights tend to run from 6.3 to nine pounds, again depending on variant type. Stocks are a standard, hand checkered, French walnut type with semi-

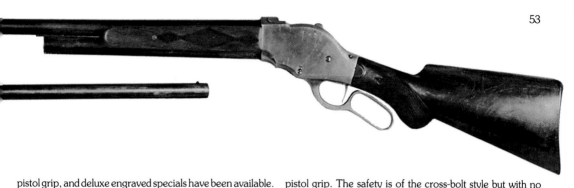

pistol grip, and deluxe engraved specials have been available.

This was the world's first successful automatic shotgun, an unheard-of thing in its day, and immediately successful, selling 10,000 guns in its first year of production. John M Browning took out four patents on this revolutionary firearm. US Patent Number 659,507, filed on 8 February 1900, was granted on 9 October 1900; US Patent Number 689,283, filed on 18 March 1901, was granted on 17 December 1901; US Patent Number 710,094, filed on 11 January 1902, was granted on 30 September 1902; and US Patent Number 812,236, filed on 30 June 1904, was granted on 13 February 1906.

The automatic shotgun was first manufactured by Fabrique Nationale, for the Browning Arms Company, in 1903. In 1905 the Remington Arms Company was licensed to manufacture and sell the gun, bringing it out as the Remington Model 11 Automatic Shotgun. The Browning Automatic-5 is the Browning Arms Company model.

The Automatic utilizes the forces which are generated by firing the cartridge to eject the empty case, to load a fresh round from the magazine into the chamber, and to cock the gun automatically. Through a simple, but highly effective, adjustable friction break and shock absorber, Browning was able to make the gun adaptable to whatever cartridge loads were desired.

The original Browning Automatic Shotgun is in 12 gauge only, with a 28-inch barrel; full, modified or cylinder bore choke; and is chambered for any cartridge up to 2.8 inches. Its weight is about 7.8 pounds and stocks tend to be English walnut with a straight grip. The gun exists in three models, all of them takedown-capable. These represent the Regular, Trap and Messenger (20-inch barrel) guns. While these models are generally of five-shot capacity, there are also two-shot variants.

The Remington model that was manufactured in 1905 is practically identical to the Browning Automatic-5, many of the parts being interchangeable. It differs in the following details. The butt plate is of hard rubber. The stock had a full

pistol grip. The safety is of the cross-bolt style but with no finger piece. There is no magazine cut-off. The forearm has a reinforcing dowel which the Automatic-5 does not have. The carrier is of the old style, without the quick-loading feature. The front trigger-plate screw is a pin, and the rear trigger-plate screw has no locking screw. On some of the early Remington models, the carrier latch screw and cartridge stop screw are pins with a transverse locking screw. There is a fiber cushion in the rear of the receiver to stop the breechblock and firing pin. On some early models, the firing pin is identical to the Automatic-5; on later models, the firing pin was changed to a straight cylinder which won't lock when the action is open. There were other very minor differences in detail, depending on the year of manufacture.

Two variations in the Browning gun were offered in 1921—a Police Special, in 12 gauge, and a short-barrel Riot Gun, in 12, 16 and 20 gauge. The Model Sportsman, a three-shot version of the Model 11, with semi-beavertail fore end, was introduced in 1931. All models of the Model 11 and the Model Sportsman were discontinued in 1948. The Remington Model 11-48 Autoloader was introduced in 1949, and except for changes made to streamline the gun and make it easier to manufacture, it still has the same basic action as its predecessor.

The Browning Automatic-5 is still in production and is offered in many different specifications. Combined production figures on this gun cannot even be estimated.

Model 17 Pump Action Shotgun (Remington). This is a pump action, underloading shotgun in 20 gauge, with a five-shot tubular magazine, and groove and bead sights. Barrel lengths for all variants tend to be 18.5, 20, 26, 28, 30 and 32 inches, with standard full choke and optional chokes. The weights for various examples tend to be around 5.3 pounds, and stocks are as follows. Standard and riot models are checkered with a straight grip and hard rubber butt plate, while police models feature a pistol grip only, with no shoulder stock.

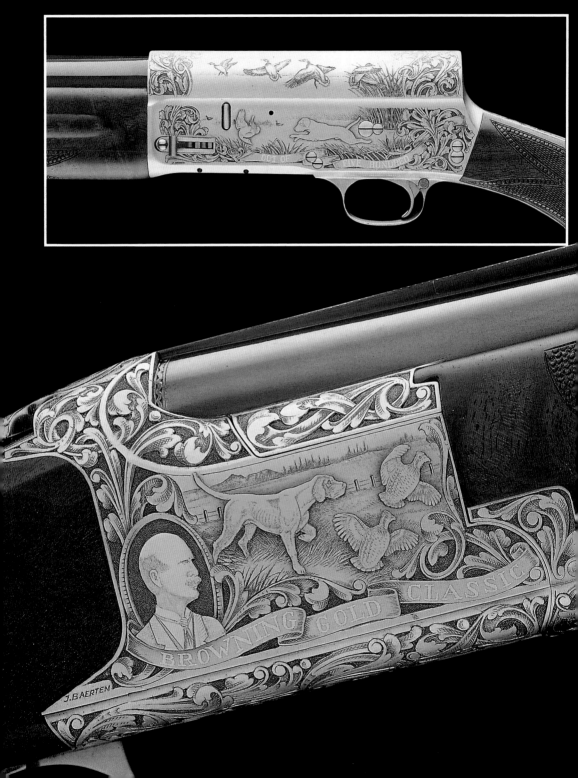

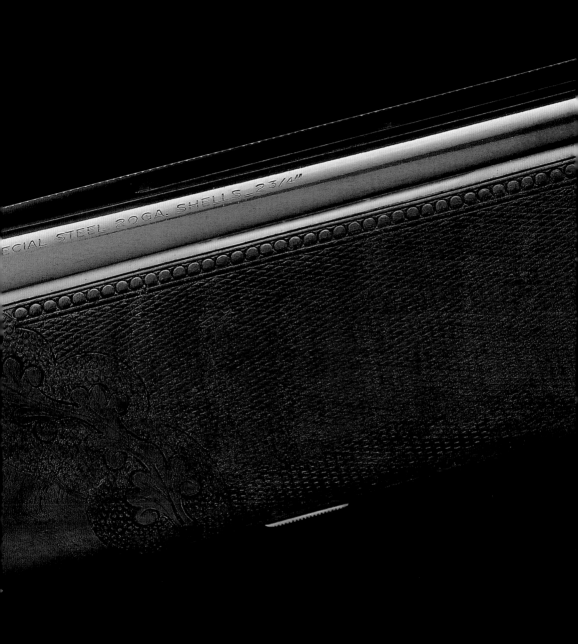

ECIAL STEEL 20 GA. SHELLS 2 3/4"

Above: This beautifully engraved Gold Classic Superposed has a finely grained walnut stock and forearm with intricate hand checkering. The right side of the receiver features a Labrador and a pair of grouse in gold inlay. *Far left:* The receiver of this Gold Classic Automatic-5 displays a Labrador fetching a downed duck and a banner with the words "001 of Five Hundred."

This was John M Browning's last repeating shotgun. The patent for this gun was filed on 26 November 1913, and US Patent Number 1,143,170 was granted on 15 June 1915. The Remington Arms Company was granted manufacturing and sales rights to this model, introducing it in 1921 as the Remington Model 17.

All production was discontinued in 1933. Since its patent expiration, the design has been used with marked success by various manufacturers, with examples being made in all gauges. For instance, the Ithaca Model 37 Pump is of the same basic design except that it is in 12 gauge. No production figures are available on the Model 17.

Superposed Shotgun (Fabrique Nationale and Browning). This is an over and under double-barrel shotgun in 12, 20, 28 and .410 gauge, including Magnum loads.

Model style: Standard, Magnum, Lightning, Lightning Trap, and Broadway Trap 12 gauge; Standard and Lightning 20 gauge; Standard 28 gauge; Standard .410 gauge. Barrel lengths for the Browning Superposed Shotgun run the gamut from 26.5 to 32 inches, with extra barrel sets available for each gun, and can be found in all possible choke borings. Weight also varies widely—from six to 8.1 pounds, depending upon varaint and barrel choice, etcetera. Stocks are a standard hand-checkered, hand-rubbed walnut, with a semi-pistol grip. However, deluxe engraved specials are also common.

The Superposed Shotgun was John M Browning's last invention. Patent applications on this gun were filed on 15 October 1923 and 29 September 1924, and US Patent Numbers 1,578, 638—39 were granted on 30 March 1926. First Produced by Fabrique Nationale in 1930, the Superposed Shotgun appeared in the Browning Arms Company line in 1931.

The barrels are mounted one above the other, rather than side by side, to permit the improved accuracy of a single sighting plane. Shells are placed directly in the two chambers. Automatic ejectors flip out the spent shells when the gun is opened. Unfired shells are merely elevated by the ejectors for easy removal by hand, if desired. This is made possible by the various physical differences between spent and unspent shells.

The first Superposed models had double triggers. Later, John M Browning's son, Val A Browning, designed twin single triggers for the gun, and ultimately the single selective trigger. The twin single triggers differed from the double triggers in that after selecting and shooting one round in one barrel, a second pull on the same trigger would fire the remaining barrel, thus eliminating the necessity of moving the finger from one trigger to the other. In the final version, the single selective trigger fires both barrels, either barrel first, by moving a thumb selector. The Superposed was initially provided only in 12 gauge in the United States. Val A Browning later designed the 20 gauge variant for American sale.

The Browning Superposed is still in production and is currently available in a wide variety of custom specifications.

At top: A Browning designed Remington Model 17 Pump Action Shotgun. *Below, at bottom:* Two of Browning's models for the automatic shotgun. Since the fourteenth century Belgian engravers have been renowned for their artistry, and the tradition lives on today at Fabrique Nationale in Liège. *Below:* The Presentation One Superposed, with a pair of gold inlaid mallard ducks on an engraved receiver, is an example of their fine craftsmanship. *Overleaf:* A Black Duck Limited Edition Superposed Shotgun and a BAR Big Game Series Limited Edition.

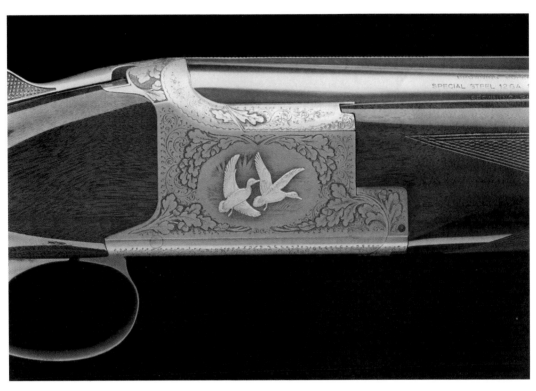

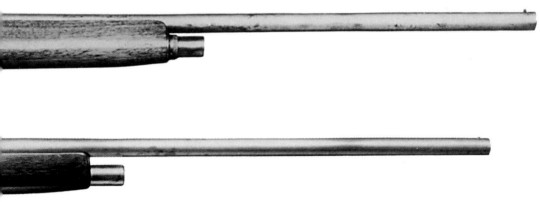

Little-Known Browning Arms

Among the patents that were assigned to John M Browning, but not specifically mentioned in the previous listings, are 32 patents covering entire firearms for which there are no known surviving models. In addition, several models were often covered by the same patent. It is known, for example, that Browning designed several additional experimental models of both the .45 caliber Government and the .22 caliber long rifle practice pistols. The total of John M Browning's original firearms models is therefore well over 100.

Also, no mention has been made here of the Browning Double Automatic. This shotgun was invented not by John M Browning himself, but by his son Val A Browning— while he was president of the Browning Arms Company. The invention of this firearm is of note here, as it is one of a number of fine examples of the continuing Browning family gunmaking tradition.

To clear up an ongoing confusion, mention should also be made of the FN-Browning Light Automatic Rifle, Caliber .308, the standard infantry rifle for the NATO nations. Despite its name, this weapon was not a Browning invention. The following excerpt from the Fabrique Nationale descriptive brochure on this gun explains why it is so designated:

'The designer of this weapon was Mr DD Saive, Chief of Weapon Design and Development at FN, who, in the course of his career was able to gain an extensive experience in automatic weapons. For many years he collaborated with the great inventor, JM Browning. It is not surprising, therefore, that one finds in this rifle, in several places, features which first appeared in Browning mechanisms (gas intake and piston, wire-spring-actuated extractor, recoil spring housed in the buttstock); and thus it can be said that the weapon is of Browning inspiration— a natural consequence of more than 50 years of continuous collaboration between the FN and Browning companies.'

Of the 44 firearms John M Browning sold to the Winchester Repeating Arms Company, thirty-one were rifles. Of this number only seven were manufactured. Among the 24 not manufactured are a .38 caliber, tubular magazine, lever action rifle; a .30 caliber, box magazine, pump action rifle; and a .30 caliber repeater in which the operant mechanism is very much like a backwards-mounted lever action.

In addition, John M Browning submitted three other rifles to Winchester which were neither patented nor manufactured. These include a .44 caliber pump action repeating rifle, a .22 caliber single shot rifle, and a .45 caliber lever action single shot rifle. It is assumed that these guns were not patented because of previously existing patents, and consequently were not purchased by Winchester, although the models remain in the Winchester Gun Museum.

Also, John M Browning sold 13 shotguns to the Winchester Repeating Arms Company. Of these, only three were manufactured. One of the shotguns John M Browning submitted to Winchester, a 12-gauge, pump action shotgun, was neither patented nor manufactured. It is assumed that this gun was not patented because of previously existing patents and consequently was not purchased by Winchester, although the model remains in the Winchester Gun Museum.

Below: John M Browning's workbench, with some tools made by his father way back in the 1830s. *Opposite:* An intricately engraved Superposed steel receiver inscribed with the engraver's name.

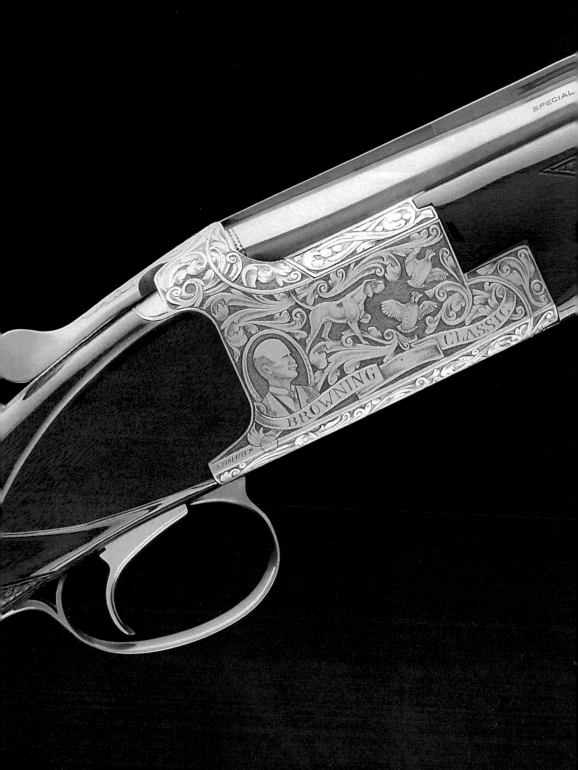

Browning's
and Aircraft

A Selection of Browning Machine Guns

In the fall of 1889, John M Browning made several experimental models which served as preliminary steps in his development of a gas-operated, fully automatic gun that would fire continuously as long as the trigger is pulled and a supply of cartridges is being fed into it.

The first of these was derived from a Winchester Model 73 lever action. Browning put a hinge on the muzzle end of the barrel which operated a gas flapper that was attached to the modified lever of the rifle. Though this crude machine gun was built in a single day, it worked. Several others were made—all working out the machinations necessary to perfect his gas-operation experiment.

A patent application was filed on 6 January 1890, and US Patent Number 417,782 was granted on 29 March 1892. This was John M Browning's first patent embodying his gas-operation principle for the machine gun. Browning noted in the application that 'This invention is applicable to machine guns and also to firearms.'

The final experiment had a concave cap with a hole in the center which was fitted directly over the muzzle. When the bullet passed down the barrel and through the hole in the cap, the expanding gases that followed the bullet forced this cap forward. Attached to the cap was a spring-loaded operating lever that was connected also to the gun's firing mechanism. When a shot was fired, the cap would move, trigger the operating lever which in turn 'pulled' the trigger, creating an identical repeat process. This could, hypothetically, go on indefinitely, providing the trigger was held in the 'pulled' position, there were enough cartridges to fire and the barrel of the weapon didn't melt from the heat of continuous firing.

This weapon is referred to in Browning's patent as an 'Automatic Magazine Gun.' It fired .44-40 black powder ammo at the rate of 960 rounds per minute, and weighed approximately eight pounds. It was not, however, considered ready for production.

The first Browning Gas-Operated Machine Gun.
This is a gas-operated full automatic weapon in .45-70 caliber, with a belt magazine, a cyclic rate of fire of 600

Below: **John M Browning's first gas-operated machine gun, circa 1890–91, made solely for testing.** *Opposite:* **Browning and an infantry variant of the Model 1917 .30 Caliber Machine Gun.**

Machine Guns
Cannon

rounds per minute and a barrel length of 22.5 inches. This gun weighs 40 pounds with its mount.

This was John M Browning's first fully developed machine gun. This gun was invented in 1890–91. The patent application was filed on 3 August 1891 and US Patent Number 471,783 was granted on 29 March 1892.

Browning placed a bracket on the muzzle end of the barrel of this model. On it, a lever was hung on a pivot, so that one end of the lever formed a cap over the front of the muzzle. This cap contained an aperture corresponding to the bore of the barrel, to allow passage of the bullet. The muzzle bracket acted as a spacer to keep the lever cap a short distance forward of the muzzle, thereby forming a small, enclosed gas chamber between the end of the barrel and the cap.

When a shot was fired, the expanding gases following the bullet pushed the cap forward. The cap lever was in turn connected to the action by a series of rods and levers—thus, the forward action of the muzzle cap was the initial impetus which carried through the mechanism, effecting extraction, ejection, feeding, loading and firing of the cartridges automatically.

Gas-Operated Breechloading Gun. The patent application on this gun was filed on 11 July 1892, and US Patent Number 502,549 was granted on 1 August 1893.

A new idea appeared in this model: the energy was not taken from the muzzle—instead, a hole was drilled through the barrel, tapping the high-pressure gases directly behind the bullet before the bullet had left the barrel. Actually, holes were drilled on both sides of the barrel, and a pair of flappers was positioned near these holes.

The flappers were, in turn, attached to the operating rod on the bottom of the barrel. The expanding gas that was tapped from behind the bullet (via the holes) caused both flappers to pivot rearward, imparting motion to the operating rod, and subsequently to the operating mechanism.

One possible reason for this arrangement is that it may have helped to ameliorate the instability at the muzzle which resulted from the unbalanced action of a single flapper. However, as soon as this two-flapper mechanism was created, Browning became aware of many possibilities for simpler mechanisms.

Model 1895 Automatic Machine Gun (Colt). This is a gas-operated machine gun in .30-40 Krag and 6mm Lee Enfield calibers, with a belt magazine and a cyclic rate of fire of 400 rounds per minute. It is an air-cooled weapon, with a 21.5-inch barrel, and weighs 40 pounds.

This was the first fully automatic weapon to be purchased by the United States Government. The first patent applica-

tion of this gun was filed 7 November 1892, and US Patent Number 544,657 was granted on 20 August 1895. Various modifications of the gun were covered by various US Patents—Number 544,658, filed on 15 March 1893 and granted 20 August 1895; Number 544,659, filed on 11 September 1893 and granted on 20 August 1895; and Number 543,567, filed on 16 April 1895 and granted 30 July 1895.

Arrangements were made with Colt's Patent Firearms Manufacturing Company in 1895 for the gun's manufacture the same year. In January 1896, the Model 95 was tested by the Navy in competitive trials. Its successful performance resulted in the Navy's placing an order with Colt for fifty of these guns. In the hands of US Marines, these Model 1895 Machine Guns saved the foreign legations in Peking during the Boxer Rebellion.

After their use in the Spanish-American War, the Model 95 acquired a nickname—the 'Browning Peacemaker.' At the outbreak of World War I, the Model 1895 comprised a large portion of the United States machine gun arsenal, and though by the second decade of the twentieth century, the gun was outdated, a number were manufactured for military use in the interim before the Model 1917 Browning Heavy Water-Cooled Machine Gun went into production. The Model 1895 was then relegated to training use.

The Model 1895 Machine Gun employed a hole drilled through the barrel near the muzzle, which powered a piston that worked the mechanism of the gun via a swinging lever. When the mechanism was activated, it fed, fired, extracted and ejected the cartridges automatically. The firing cycle was continuous as long as the trigger was depressed and ammunition was supplied, and a very heavy barrel was used on the theory that the extra metal would help prevent heat buildup. The unusual movement of the piston, which swung in a half arc beneath the barrel, gave the gun an additional nickname—'The Potato Digger.'

This gun was discontinued in 1917. Total production figures are not available. Some 1500 were produced during World War I.

Experimental Gas-Operated Firearm. This is an experimental gas-operated machine gun in .44 caliber, with a 20-shot box magazine and a cyclic rate of fire of 720 rounds per minute. It has a 14.5-inch barrel and weighs 7.3 pounds.

This was an important test model, designed to improve upon the ideas that were already in operation on the Model 1895 Machine Gun. After their testing in the Experimental Gas-Operated Firearm, many of these improvements would go into the later models that would see action in some of the most intensive conflicts in our present war-heavy century.

Gas-Operated Breechloading Gun
.44 Caliber Experimental Model

Model 1917 .30 Caliber

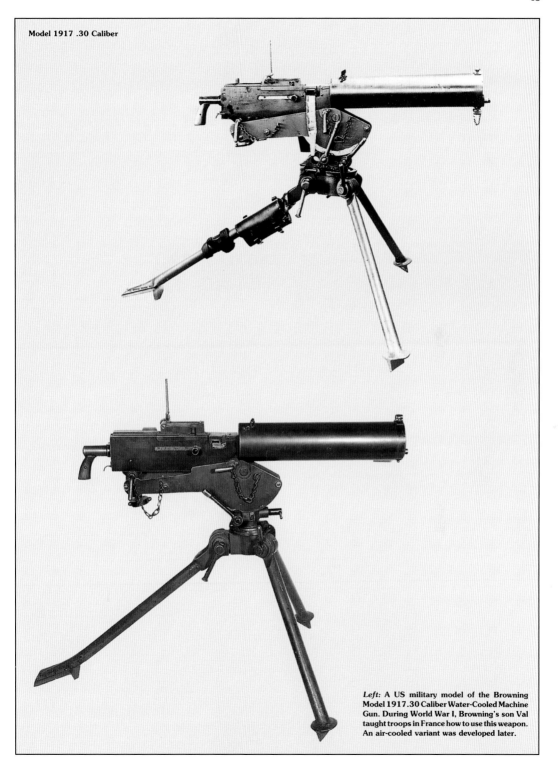

Left: A US military model of the Browning Model 1917 .30 Caliber Water-Cooled Machine Gun. During World War I, Browning's son Val taught troops in France how to use this weapon. An air-cooled variant was developed later.

The patent application on this gun was filed 8 December 1894, and US Patent Number 544,661 was granted on 20 August 1895. Because of its hand-forged parts and unfinished appearance, this experimental model is often erroneously assumed to be a very early Browning gas-operated gun.

However, Browning's main purpose in designing it was to effect an improvement in the gas port, which, in all his prior machine guns (including the Colt Model 1895) had undergone little change. As stated in the patent, 'The objects of the invention being to avoid the fouling and clogging of the mechanism by the gases and to prevent the escape of the gases until after the lever shall have commenced its opening movement and received its initial force, and to prevent the lateral spread of the gases and to generally improve and simplify the construction of the gas operated mechanism.'

One very important idea that is present in this experimental gun centers on an alternate method of constructing the gas port. This method entails putting an elbow on the gas vent, so that the force of the gases are applied along the axis of the barrel—enabling the use of a piston to operate the mechanism rather than a swinging arm. This is an idea that has seen much usage since.

Model 1917 .30 Caliber Machine Gun (Colt, Remington, Westinghouse and Others). This is a short recoil-operated, water-cooled and later air-cooled, fully automatic machine gun in .30-06 caliber with a link belt magazine and a 20-inch barrel. Its cyclic rate of fire is 600 rounds per minute (with adjustments in later variants), and it weighs 37 pounds with its water jacket filled, and 22 pounds in air-cooled versions.

Versions of this gun have made history in World Wars I and II, and the Korean War. Until recently, it continued to

Left: John M Browning testing one of his machine guns. *Below:* A machine gunbelt loading device, one of Browning's miscellaneous inventions, designed in 1899 at the request of US Army Ordance.

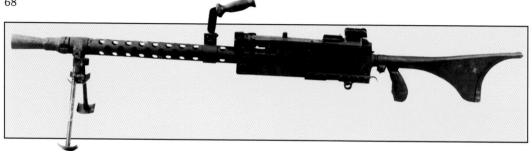

occupy a prominent place in the military arsenal of the United States. All this began back in 1900, when John M Browning invented a machine gun to replace the outdated Model 1895. The patent application for this new gun was filed on 19 June 1900, and US Patent Number 768,934 was granted on 23 July 1901.

Although sometimes referred to as the Browning Model 1901, the design of 1900 was never manufactured, since the government lacked interest in military weapons at the time. Nevertheless, its basic operating features are the same as those of all Browning machine guns since produced.

Browning continued working on the gun intermittently over the years, modifying its mechanism to eject from the bottom rather than the right side, and increasing its rate of fire. The patent application covering these improvements was filed on 3 October 1916, and US Patent Number 1,293,021 was granted on 4 February 1919.

The model embodying these changes—known as the Browning .30 Caliber Heavy Machine Gun—was first publicly demonstrated at Congress Heights, Washington, DC, on 27 February 1917. In May 1917, official tests were held at

Springfield Armory. Following the tests, a board appointed by the US Secretary of War recommended its immediate adoption. The first combat use of the Model 1917 was by the 79th Division in France on 26 September 1918.

In the short-recoil system, the barrel and breechblock are locked together when the gun is fired and recoil together for a short distance, until the bullet clears the barrel and the gas pressure diminishes, at which point they unlock, and the breechblock alone continues to recoil. During this time, all this energy compresses the mechanism's springs, so that, as they become unsprung, the springs return all parts to their 'battery,' or firing positions. During the recoil of the breechblock, the fired cartridge is extracted from the barrel and ejected, and a fresh cartridge is fed into the chamber and fired. The firing cycle is continuous as long as the trigger is depressed and a supply of ammunition available.

This model was water-cooled, but Browning also made a later air-cooled variant. This air-cooled weapon was the first

At top: The .30 Caliber Air-Cooled Machine Gun was used by American forces during World War I, World War II and the Korean War. *Left:* US soldiers fire the same gun in combat during World War II.

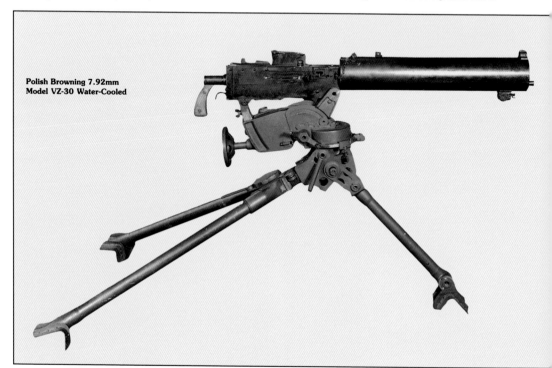

Polish Browning 7.92mm
Model VZ-30 Water-Cooled

.30 Caliber Air-Cooled

Browning .50 caliber machine guns were standard armament for World War II Allied fighters like this P-51 Mustang *(below)*. *At right* is a section view of same, showing its control cockpit detail. *Far right:* Armorers of the 45th Fighter Squadron, 15th Fighter Group replenish a P-51's guns.

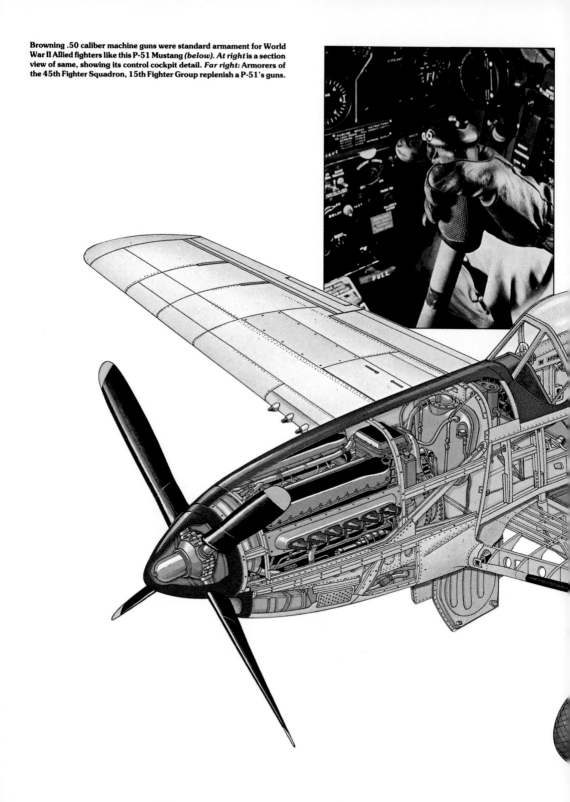

Above: A Browning .50 caliber M2 on a halftrack in Rome 1944. Here it is being used for air defense. *Right:* A waist gunner aboard a B-17F at Bovington, England readies his weapon, a Browning .50 Caliber Machine Gun.

in the US to be successfully used aboard pursuit planes—the pilot looked along his sights and aimed at the target by maneuvering his ship. To enable the use of a forward-mounted machine gun on a fighter plane without literally shooting one's own propeller off, the firing mechanism of the gun was synchronized with the motor of the plane so that the bullets passed through the spaces between the revolving propeller's blades.

The .30 caliber air-cooled Browning, in its diverse con-figurations, was one of America's most important military weapons in World War II. Many modified versions of this gun have appeared since its invention. It generally weighs approximately 22 pounds, and has an increased rate of fire of approximately 700 rounds per minute, although some aircraft models were stepped up to as high as 1300 rounds per minute for duty in the superheated skies of the 1940s and early 1950s.

The US Government contracted with three companies for World War I production of the Model 1917. Between its official adoption in 1917 and the Armistice of the following year, nearly 43,000 were produced: Westinghouse accounted for 30,150, Remington, 12,000 and Colt 600. Over a million units of the World War II Browning .30M2 were manufactured. Most countries have since acquired similar weapons. Total production can only be estimated as well into the millions.

Browning Water-Cooled Machine Gun in .50 Caliber (Colt and others). This is a short recoil-operated, water-cooled and later air-cooled, machine gun in .50 caliber with a link belt magazine and a 39-inch barrel. This weapon weighs 82 pounds with a full water jacket—air-cooled variants weigh less.

To meet the increased threat of armored combat vehicles in World War I, General John 'Blackjack' Pershing, Commander of the American Expeditionary Forces, requested a machine gun cartridge that was heavier and more powerful than the .30 caliber cartridge then in use by the AEF.

A .50-caliber cartridge was developed—it hurled a 1.8 ounce bullet to a muzzle velocity of 2750 feet per second. To shoot this powerful load, John M Browning developed

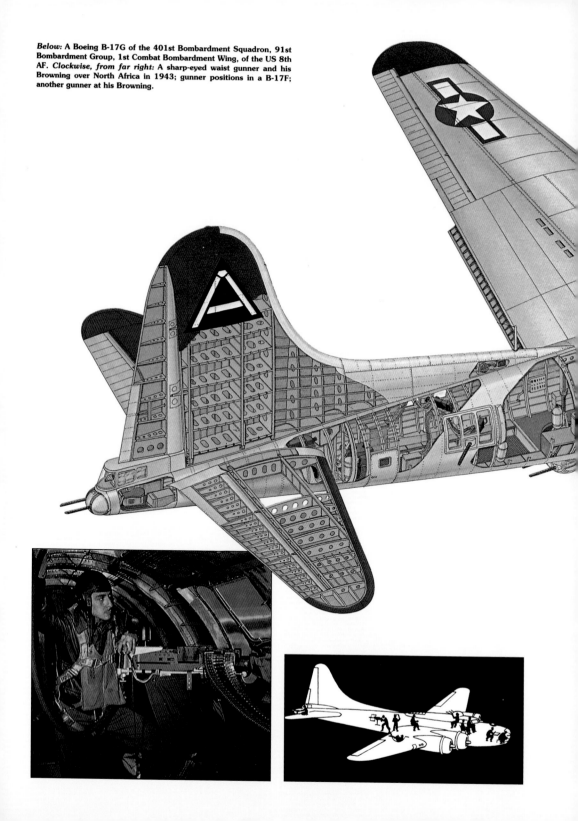

Below: A Boeing B-17G of the 401st Bombardment Squadron, 91st Bombardment Group, 1st Combat Bombardment Wing, of the US 8th AF. *Clockwise, from far right:* A sharp-eyed waist gunner and his Browning over North Africa in 1943; gunner positions in a B-17F; another gunner at his Browning.

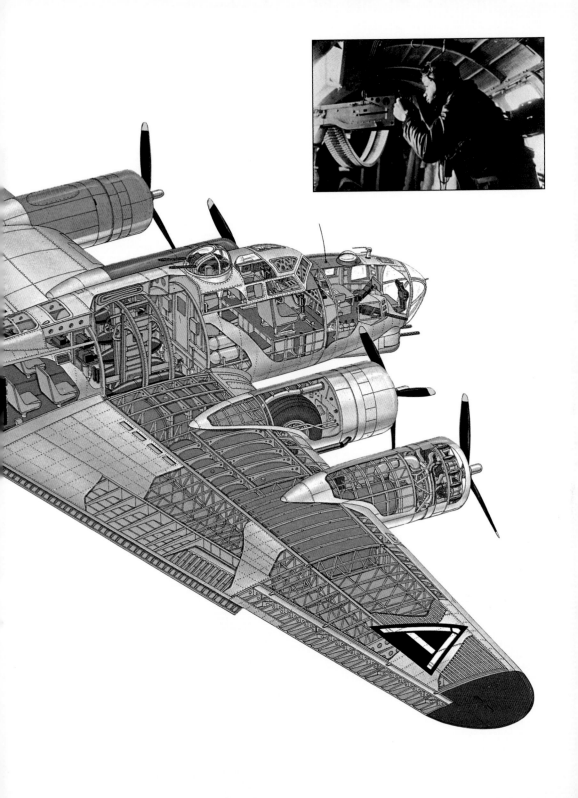

the Browning .50 Caliber Water-Cooled Machine Gun. He began work in July of 1917, and the gun was first test-fired, in Colt's meadow in Hartford, Connecticut, a little over a year later. The patent application for the gun was filed on 31 July 1923, and US Patent Number 1,628,226 was granted on 10 May 1927.

Developed too late to see use in World War I, the Browning .50-Caliber Machine Gun played a prominent role in World War II and the Korean War. This gun has the same basic operating features as the .30 caliber, but through the use of a unique oil buffer the necessary strengthening of the gun is accomplished without a proportional increase in weight. The oil buffer absorbs excess recoil energy, thereby effectively reducing undue strain on the parts. In addition, this oil buffer provides a means of controlling the rate of fire.

Browning also incorporated double spade grips on the .50 caliber, as these gave a two-handed—and thus more stable—control of the gun than did the pistol grip used on the smaller .30 caliber. To date, no less than 66 known models of the Browning recoil-operated machine gun have been manufactured by the US and various Allied countries. The majority of these have been .50 caliber. Other highly effective models which followed the original prototype were the Water-Cooled Infantry, the Water-Cooled Anti-Aircraft Single and Twin Mount, and the Air-Cooled Tank Gun. One of the later aircraft models, the .50M3, was stepped up to a cyclic rate of fire of 1200 rounds per minute.

During World War II, the .50M2 was produced by the following companies: Colt's Patent Firearms Manufacturing Company, High Standard Company, Savage Arms Corporation, Buffalo Arms Corporation, Frigidaire, AC Spark Plug, Brown-Lipe-Chappin, Saginaw Division of General Motors Corporation and Kelsey-Hayes Wheel Company. Of the 3,283,837 Browning machine guns produced in this country during World War II, approximately two million were .50 calibers. Like the Browning .30 caliber, this gun has been widely copied by other countries, and total production can only be estimated as well into the millions.

Necessarily, not all of John M Browning's machine gun and fully automatic weapon mechanisms can be covered here—nor can the work of such important modifiers of original Browning designs as Fred Moore and Colonel S Gordon Green. For those seeking further information, please consult *Volume One of The Machine Gun: History, Evolution and Development of Manual, Automatic and Airborne Repeating Weapons,* by George M Chinn, Lt Col, USMC Retired, published by the Superintendent of Documents, Washington, DC.

John M Browning's Aircraft Cannon

Browning 37mm Aircraft Cannon (Colt, Vickers and others). This is a long recoil-operated aircraft cannon of 37mm caliber with a barrel length of 44 inches and a cyclic rate of fire of 135 rounds per minute. This cannon weighs approximately 313 pounds without magazine attached, and approximately 406 pounds with a 15-shot magazine attached. These specifications are based on the M4 model.

John M Browning began work on his first aircraft cannon in early 1921; three months later, it was successfully test-fired in the hills outside of Ogden. The patent applications for this gun were filed on 15 December 1923, 11 April 1924 and 28 April 1924, and US Patent Numbers 1,525,065—67 were granted on 3 February 1925.

The cannon was successfully demonstrated to US Army officials at the Aberdeen Proving Ground in mid-1921, when it fired a series of one-pound projectiles at a muzzle velocity of 1400 feet per second, and a cyclical rate of 150 rounds per minute. Shortly after, Browning designed two additional models, each firing heavier projectiles—the first at 2000 feet per second, and the second at approximately 3000 feet per second.

In 1929, a small number of these weapons were manufactured by the Vickers Arms Company in England for sale to Spain. The US Government, for a time apathetic to the production of new military weapons, did not renew interest in the gun until 1935, when the Army Air Corps model of Browning's cannon was produced by request. This cannon was entered on the ordnance lists as the M4.

The World War II model, the M9, was used in rather limited quantities by the US Army Air Corps, which eventually decided that such heavy armament was neither practical nor essential. This cannon did, however, see much use in the hands of the Russians. During their most critical defensive combat with Germany, the M9 was their primary aerial cannon. Several thousand of these guns—and the P-39 Bell Aircobra planes on which they were mounted—had been sent to Russia by the US Government. The M9s' high-velocity, armor-piercing projectiles proved to be quite effective against German tanks.

The basis of the long-recoil principle is that both recoil and counter-recoil are controlled by a hydro-spring buffing mechanism. The breechblock is of the vertical, sliding-wedge type. When the projectile is fired and driven down the bore of the barrel, the barrel, breechblock, and locking frame, all locked together, recoil rearward 10 inches before the breechblock cams downward, cocking the hammer, ejecting the empty case, and loading a new shell—at which point, the mechanism locks up and is again ready for firing.

The M9 was principally intended for use by aircraft. It is either mounted to fire through the hub of the propeller or from the wings. With little change, it can be fed from either right or left. The muzzle velocity of its standard round is 3050 feet per second. The total production of all Browning Aircraft Cannons, including World War II, is under 100,000 units.

The United States Government sent several thousand P-39 Bell Aircobra planes mounted with Browning M9s to Russia. With their armor-piercing projectiles, the M9s were a crucial element in the Russians' aerial battles against the Germans. *Above:* A Bell P39D-1 in flight.

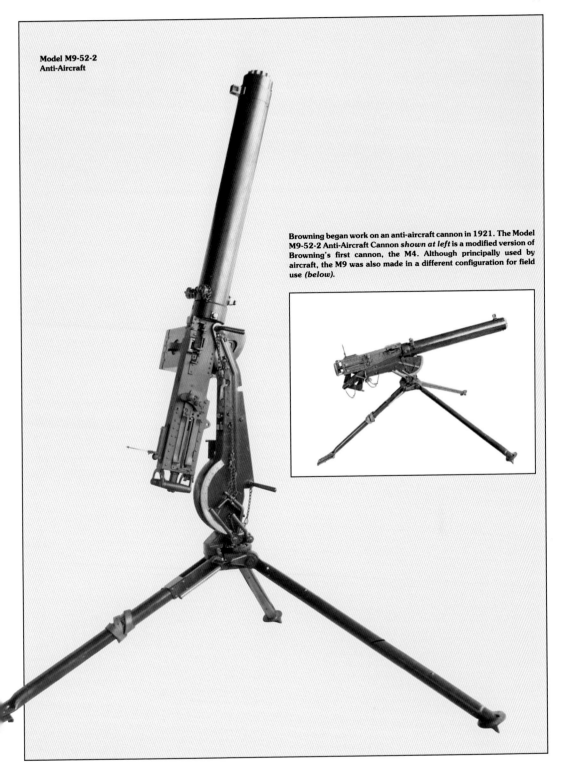

**Model M9-52-2
Anti-Aircraft**

Browning began work on an anti-aircraft cannon in 1921. The Model M9-52-2 Anti-Aircraft Cannon *shown at left* is a modified version of Browning's first cannon, the M4. Although principally used by aircraft, the M9 was also made in a different configuration for field use *(below)*.

Fabrique

The History

It is quite natural that firearms enthusiasts throughout the world should regard Liège, Belgium as their capital; it has been steeped in gunmaking tradition for nearly six centuries. Indeed, Fabrique Nationale, the historical heir of a firearm manufacturing tradition dating from the fourteenth century has its home in the suburbs of this gracious city of the Meuse Valley.

Way back in the 1500s, metal workers grouped under the name of 'Bon Metier Des Febvres' (the 'Good Guild of Metal Workers') had a central role in the industrial and economic life of Liège. These men were totally dedicated to the manufacturing of guns and military weapons, and to the making of the highest quality gunpowder and projectiles.

In this same era, the Princes-Evaques managed to maintain the Principality of Liège in a state of political neutrality, and made the city's fortune by promoting its weapons to a number of warring neighbor-states.

In the seventeenth century the talent of metal workers from Liège came to be recognized throughout Europe. During this period a Swedish a banker named Louis de Geer financed a dozen iron mills, nail works, gun foundries and arms factories. Many of the workers at these various facilities were workers from the Principality of Liège.

At the Court of the French King Louis XIV, richly adorned firearms were in vogue, and the ability to create a finely-crafted, expensively ornate firearm had become prized as one of the highest arts—and was rewarded richly by the extravagant French royalty. Therefore attracted to the French Court, the great master armorers of the time left their own countries and settled in northwestern France. And from Liège came Adrien Reynier, who became the personal gunsmith to Louis XIV. The guns and pistols of Reynier and his fellow countrymen became renowned as masterpieces of classical arm manufacturing.

In 1672, a proclamation was issued in Liège that required arms makers to have their products tested and hallmarked. Fabrique Nationale's concern for quality and perfection therefore has its roots in that long-ago proclamation, and it is still a source of pride that every FN firearm is thoroughly proven to a very high standard of excellence.

In the eighteenth century, the export of firearms was an ever-increasing source of prosperity for Liège. By 1750, annual firearms production in Liège had reached 100,000 weapons. By 1788, the 70 to 80 arms manufacturers in Liège were exporting more than 200,000 weapons annually at a value of three or four million guilders. And the quality of even the lowest-grade firearms was steadily improving, for, due to the quality demands made by the European aristocracy, increasingly sophisticated firearms manufacturing techniques were steadily being invented. The French Revolution and its aftermath were disastrous for the gunmakers of Liège, and the number of active workshops soon fell to 14, and the government Proof House, where final inspection of firearms took place, fell into disuse. Since French royalty no longer existed, the demand for extremely high-quality firearms had dropped to almost nil. Most orders were now for only the most utilitarian of weapons.

In 1810, an imperial decree reestablished the Proof House, and at about that same time, innovation began to trickle back into the shops, for a production line was set up whereby separately machined parts could be fully assembled after only very small adjustments.

Under the government of Napoleon Bonaparte, tastes in firearms once again turned toward the higher grades, and

In 1984, Browning introduced the Classic and Gold Classic series to commemorate three of John Browning's greatest inventions. Right, from top to bottom: The Classic Automatic-5 Shotgun, the Classic Over & Under Shotgun and the Classic 9mm Hi-Power Pistol.

Nationale

technical improvements became the watch word of the day. More than ever before, the gunmakers of Liège were asked to make vast quantities of arms for their Parisian customers.

The 19th century may be called the Golden Age of gunsmithing in Liège, and was marked by significant advances in the craft. A whole range of inventions had contributed to revolutionizing the design and the manufacture of military firearms. In particular, many of these inventions were derived from the percussion-cap principle that was patented by Alexander Forsyth in 1807.

Heretofore, ignition of the powder charge in a firearm was effected by one of several methods—all of which had their own mechanism of application—the wick, or 'match'; the spark-generating friction wheel; and the spark-striking chip of flint. A percussion cap is a small brass or copper cup that contains a small amount of fulminate of mercury, which explodes when struck by a blow. The percussion cap rests on a nipple which is tapped into the the breech of the gun. When it is struck by the firearm's hammer, the charge in the cap expends its energy through the hole in the nipple, igniting the powder charge in the gun's chamber. This was so much more efficient than the wheellock, flintlock or matchlock that its acceptance was rapid and widespread.

Soon enough, experiments with self-contained cartridges of various types resulted in cardboard cartridges made for guns with long, needle-like firing pins. These were inefficient, as the cartridges tended to swell and stick in the chamber. At about the same time, the idea of using paper cartridges in conjunction with percussion caps was leading to the design of many of the weapons that were used in the American Civil War. Shortly after this, however, the percussion cap was made an integral part of a limited-expansion brass cartridge, and the basic mechanism of the modern firearm was formed around this combination.

The progress that was made in the development of the modern cartridge also led to the development of firearms able to fire a varying number of shots. So it was that Collier's revolver appeared in about 1820; and 1835 saw the invention of Colt's revolver; Spencer's repeating rifle appeared in 1860 and Mauser's rifle appeared in 1864. These were followed by the automatic weapons inventions of the late 1880s and early 1890s. Liège's manufacturers adopted these various improvements as soon as they were introduced. In 1870, an idea which later would lead to the creation of Fabrique Nationale began to take shape. Fearing foreign competition, some manufacturers decided to unite. First they established a communal workshop—which was known as the 'Petit Syndicat,' or 'Little Syndicate'—where they made guns for the Belgian Guard.

In 1886, a partnership grouped seven manufacturers under the name of 'Les Fabricants d'Armes Reunis,' or 'The United Arms Makers.' In 1888, following an order for 150,000 repeating rifles from the Belgian Government, the Fabricants d'Armes Reunis, needing help to fulfill the order, induced their direct competitors to become their partners for the sake of fulfilling the government order. Strictly for that purpose, a limited company—actually, more of a coalition—was formed.

The name of the new Company was chosen on 8 December 1888, and it was 'Fabrique Nationale d'Armes de Guerre.' On 1 February 1890, a plot of land was purchased in the territory of Herstal, a commune lying near Liège. Today, the main workshops of Fabrique Nationale Herstal (as the firm is now formally known) firearms division are located on that very plot of ground.

The 'coalition' worked so well that the arms makers decided to keep it going. At the end of the nineteenth century, FN was exploring two new fields: the manufacturing of dual-purpose guns that could be used for both game and target shooting, and the making of bicycles. In 1896 the company board of directors approved the manufacturing of 50,000 .22 caliber sporting rifles, and in 1898, the company invented a bicycle without a chain, the so-called chainless bicycle, a version of which was immediately put into production.

Fabrique Nationale Air-Cooled Machine Gun

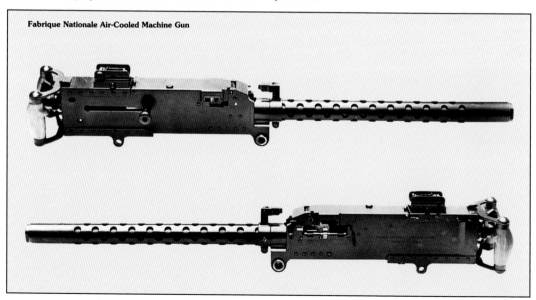

The Browning Era

In 1897, FN's Board of Directors sent its commercial director to the United States to obtain information about the latest improvements in bicycles. During his time in the US, he became acquainted with John Moses Browning and Matthew Browning. John had just applied for a patent covering his Automatic Pistol in .32 Caliber.

Browning offered the license for this pistol to FN who, upon accepting the offer, produced the first pistols of this type in 1899. The superiority of this weapon was soon widely recognized, and in 1900 it was adopted as an official sidearm for the Belgian Army Corps. This was an historic moment—it was the first time that an automatic pistol was accepted into the equipment roster of any nation's regular army.

This was just the beginning of a long and fruitful relationship. In 1902 John M Browning arrived in Herstal with the prototype of an automatic shotgun that he had previously offered, in vain, to American manufacturers. Since FN had had little success in the hunting firearms market, and especially since the .32 semiautomatic had gone over so well, the firm took an immediate interest in this weapon, and a license contract was signed. On the same day, Browning ordered 10,000 shotguns from this license, for sale under his own newly-created 'Browning Firearms Company.'

Since then, FN has manufactured about three million automatic shotguns. The substance of the relationship still maintained by FN with the Browning Company is exemplified in that long-ago agreement. Shortly after selling them the automatic shotgun license, John M Browning acceded to FN's wishes to use his name as a trademark. By granting FN the right to do so, he sealed the consolidation of interests which has, ever since, united the Browning Arms Company and Fabrique National.

As time goes on, one can judge the considerable and far-reaching consequences of that agreement. Now the FN-Browning brand name which is engraved on each firearm that is produced in by the united companies is an assurance, to shooters throughout the world, of perfection in gunmaking.

That maker's mark represents the linking of the centuries-old tradition of Liège's armorers with the inventive genius of the Browning family. Today, FN actually owns the Browning Arms Company, of which Browning Arms North America is an independent subsidiary. The tradition, firmly unified, continues.

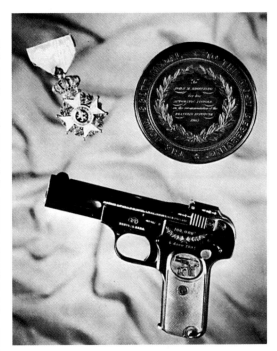

Above, clockwise from upper left: Browning's awards—The Cross of Knighthood of the Order of Leopold, The John Scott Legacy Medal and the gold-inlaid 100,000th FN Model 1900. *Left:* The Liège shop, where engravers like the one *below* ply their trade.

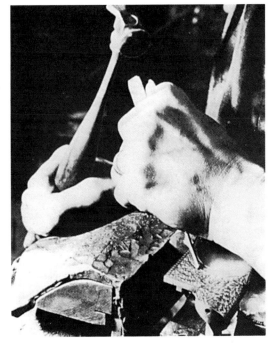

Newer
Pistols

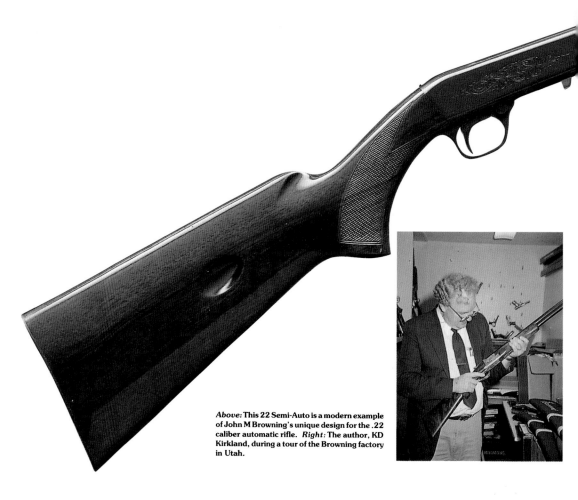

Above: This 22 Semi-Auto is a modern example of John M Browning's unique design for the .22 caliber automatic rifle. *Right:* The author, KD Kirkland, during a tour of the Browning factory in Utah.

Browning and Rifles

Browning Pistols

The following major modern Browning pistols have been manufactured for the Browning Company of Morgan, Utah by Fabrique Nationale d'Armes de Guerre (now Fabrique Nationale Herstal) of Herstal, Belgium; Arms Technology Inc of Salt Lake City, Utah; and by JP Sauer & Sohn of Eckenforde, West Germany. Many of these are either updates of or follow-ons to, original John M Browning designs.

Browning .25 Caliber Automatic Pistol. This is a blowback-operated semiautomatic pistol in .25 ACP and 6.35mm Browning caliber, with a six-shot pistol-grip magazine and a 2.1-inch barrel. This pistol weighs approximately seven ounces, and has standard hard rubber grips, with optional Nacrolac pearl grips having been available. It was manufactured from 1955 to 1969, and is based on the John M Browning design.

Browning .380 Caliber Automatic Pistol. This is a blowback operated, hammerless, semiautomatic pistol in .380 ACP caliber, with a six-shot pistol grip magazine. This pistol has two main variants, the 1955 model and the 1971 model—the former is equipped with fixed sights and a four-inch barrel, and the latter is equipped with adjustable rear/fixed front sights and a 4.4-inch barrel. The 1971 models have plastic thumbrest stocks. The earlier variants weigh 1.3 pounds, and the later variants weigh 1.4 pounds. Production

on the 1955 model ended in 1969, and production on the 1971 model ended in 1975. This is essentially an update of the original John M Browning design.

Browning Hi-Power 9mm Automatic Pistol. This is a short recoil-operated, locked breech, semiautomatic pistol in 9mm Parabellum caliber, with a 13-shot pistol grip magazine and blade front/fixed rear sights (optional adjustable rear/ramp front sights are available). This pistol has a 4.6-inch barrel and weighs 2.2 pounds. The grips are checkered walnut or Nacrolac pearl. This pistol has been manufactured from 1955 to date, and is essentially an update of the original production model of the John M Browning design.

Browning Nomad Automatic Pistol. This is a semiautomatic pistol in .22 long rifle caliber, with a pistol grip 10-shot magazine, and barrel lengths of 4.5 and 6.8 inches, with removable blade front/adjustable rear sights. This pistol weighs 2.1 pounds with the 4.5-inch barrel. Plastic stocks are standard on this pistol, which was made from from 1962 to 1974.

Browning Challenger Automatic Pistol. This is a semiautomatic pistol in .22 long rifle, with a ten-shot pistol grip magazine, and removable blade front/adjustable rear sights. Barrel lengths for this pistol range from 4.5 to 6.8 inches, and weighs 2.4 pounds with the latter barrel. This pistol has standard checkered walnut stocks, with finely figured and carved stocks on extra-fine Gold and Renaissance Models. The Challenger was manufactured from 1962 to 1975.

Browning Medalist Automatic Target Pistol. This is a semiautomatic pistol in .22 long rifle caliber, having a 10-shot magazine. Its 6.8-inch barrel has a ventilated rib, and removable blade front/click-adjustable mocrometer rear sights. The pistol weighs 2.9 pounds, and was made from 1962 to 1975.

Browning BDA Double Action Automatic Pistol. This is a locked breech, double action, semiautomatic pistol in 9mm Luger, .38 Super Auto and .45 ACP with a seven to nine-shot magazine (depending on caliber used) and a 4.4-inch barrel. This pistol weighs 1.8 pounds, has fixed sights and plastic grips, and was introduced in 1977.

Browning Rifles

The following are major modern Browning rifles that have been manufactured for Browning of Morgan, Utah, by Fabrique Nationale d'Armes de Guerre (now Fabrique Nationale Herstal) of Herstal, Belgium; Miroku Firearms Mfg Co of Tokyo, Japan; and Oy Sako Ab of Riihimaki, Finland. Many of these are either updates of, or follow-ons to, original John M Browning designs.

Browning .22 Automatic Rifle. This is an autoloading rifle in .22 short and long rifle caliber, with a 15 (short) or 10 (long rifle)-shot tubular magzine in the buttstock and a 19.3 to 24-inch barrel. It weighs approximately five pounds,

has open rear/bead front sights and pistol-grip style checkered stock. This rifle has been made from 1965 to date.

Browning High-Power Bolt Action Rifle. This is a bolt action rifle in .270 Winchester, .30-06, 7mm Remington Magnum, .300 H&H Magnum, .300 Winchester Magnum, .308 Norma Magnum, .338 Winchester Magnum, .375 H&H Magnum and .458 Winchester Magnum calibers, with a four to six shot (depending on caliber) magazine and adjustable sights and barrel lengths from 22 to 24 inches. The stock is standard with checkering and pistol grip, Monte Carlo cheekpiece, sling swivels and recoil pad on magnum models. This rifle was made from 1959 to 1974.

Browning High-Power Bolt Action Rifle, Short Action. This is a bolt action rifle (featuring a shortened, light-caliber only version of the standard bolt action) in .222 Remington and .222 Remington Magnum calibers, with a six shot magazine and a 22 to 24-inch barrel, and no factory sights included. This rifle was made from 1963 to 1974.

Browning High-Power Bolt Action Rifle, Medium Action. This is a bolt action rifle (featuring an intermediate-caliber action) in .22-250, .243 Winchester, .264 Winchester Magnum, .284 Winchester Magnum and .308 Winchester calibers, with barrel lengths that range from 22 to 24 inches. This rifle was made from 1963 to 1974.

Browning T-Bolt .22 Repeating Rifle, T-1. This is a straight-pull bolt action rifle in .22 long rifle caliber, with a five-shot clip magazine and peep rear/ramp front sights. It has a 24-inch barrel, weighs six pounds and has a plain

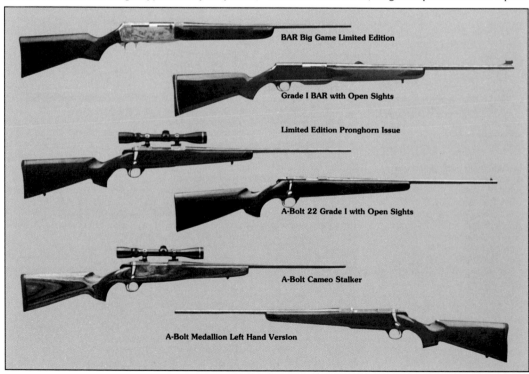

BAR Big Game Limited Edition

Grade I BAR with Open Sights

Limited Edition Pronghorn Issue

A-Bolt 22 Grade I with Open Sights

A-Bolt Cameo Stalker

A-Bolt Medallion Left Hand Version

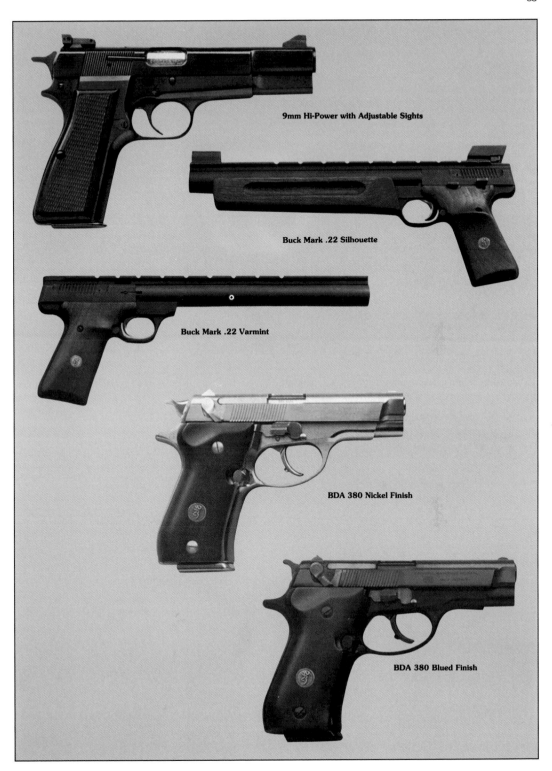

9mm Hi-Power with Adjustable Sights

Buck Mark .22 Silhouette

Buck Mark .22 Varmint

BDA 380 Nickel Finish

BDA 380 Blued Finish

The Browning Automatic Rifle was US standard military issue in the two World Wars and the Korean War and is now made for civilian use in .270 Winchester to .30-06 calibers. Not only do these guns live up to their reputation for reliability and dependability, they also make a handsome addition to any collection. *From top to bottom:* BAR Big Game Series Limited Edition, Grade IV, Grade III and Grade I.

walnut stock with pistol grip. Also available in a left hand model, it was made from 1965 to 1974.

Browning BAR Automatic Rifle. This is a gas-operated semiautomatic rifle in .243 Winchester, .270 Winchester, .308 Winchester and .30-06 calibers, with a four-shot box magazine and folding-leaf rear/hooded ramp front sights and a 22-inch barrel. It weighs about 7.5 pounds and has a checkered French walnut stock that is equipped with sling swivels, and has been manufactured from 1967 to date.

Browning BAR Magnum. This is a gas-operated semiautomatic rifle in .243 Winchester, .270 Winchester, .308 Winchester, .30-06, 7mm Remington Magnum and .300 Winchester Magnum calibers, with a four-shot (three-shot for Magnums) box magazine and folding-leaf rear/hooded ramp front sights and a 24-inch barrel. It weighs about 7.5 pounds and has a checkered French walnut stock that is equipped with sling swivels and a recoil pad. This rifle has been manufactured from 1969 to date.

Browning BL-22 Lever Action Repeating Rifle. This is a 'short-throw' lever action rifle in .22 short, long and long rifle calibers, having a 15 to 22-shot (depending on ammo used) tubular magazine and folding-leaf rear/bead fronts sights with scope mounting grooves on the receiver. It has a 20-inch barrel and weighs five pounds. Its walnut straight-grip stock is furnished with a barrel band. This rifle has been manufactured from 1970 to date.

Browning BLR Lever Action Repeating Rifle. This is a lever action repeating rifle in .243 Winchester, .308 Winchester and .358 Winchester calibers, having a four-round detachable box magazine, adjustable rear/hooded ramp front sights and a 20-inch barrel. It weighs approximately seven pounds and has a checkered walnut straight-grip stock complete with barrel band and recoil pad. This rifle has been made from 1971 to date.

Browning 78 Single Shot Rifle. This is a single shot, falling block rifle in .22-250, 6mm Remington, .243 Winchester, .25-06, 7mm Remington Magnum, .30-06 and .45-70 Government, having open rear/blade front sights on the .45-70 model only—all others have no factory sights. Its barrel length is 24 to 26 inches, with octagon or heavy round styles available, and weighs from 7.8 to 8.8 pounds, depending on caliber and barrel choice. This rifle has a fancy checkered walnut stock; more precisely, the .45-70 model has a straight-grip stock with curved butt plate, while the others have stocks equipped with a Monte Carlo comb, a cheekpiece, a pistol grip with cap and a recoil pad. The Browning 78 Single Shot Rifle has been manufactured from 1973 to date.

Browning offered a limited-edition Bicentennial 78 Set, which included a Model 78 in .45-70 caliber, with receiver engravings that featured a bison and an eagle, plus scroll engraving on top of the receiver, lever, both ends of barrel and butt plate; and a high grade walnut stock and forearm. Also as part of the set, an engraved hunting knife and stainless steel commemorative medallion, were included. Rifle, hunting knife and medallion fit in a custom alderwood presentation case, and the items in each set have matching serial numbers beginning with '1776' and ending in num-

bers from 1 to 1000. The Bicentennial 78 edition was limited to 1000 sets, and all were made in 1976.

Browning BAR .22 Automatic Rifle. This is a semiautomatic rifle in .22 long rifle caliber, with a 15-shot tubular magazine and a 20.3-inch barrel. It weighs approximately 6.3 pounds and has folding leaf rear/gold bead ramp front sights, with scope mount grooves in the receiver, and a checkered French walnut pistol-grip stock. This rifle was introduced in 1977.

Browning BPR .22 Pump Rifle. This is a hammerless slide-action repeating rifle in .22 long rifle and .22 Magnum rimfire, having a tubular magazine that holds 15 of the former or 11 of the latter ammunition, and has a 20.3-inch barrel. It weighs approximately 6.3 pounds and has folding leaf rear/gold bead ramp front sights, with scope mount grooves in the receiver, and a checkered French walnut pistol-grip stock. The Browning BPR .22 Pump was introduced in 1977.

Below: **These finely finished A-Bolt Centerfire Rifles were designed with the hunter in mind. Because the A-Bolt's 60 degree bolt requires less movement than a 90 degree bolt, the hunter can get back on target quicker for fast follow-up shots. The A-Bolt is the lightest (about seven pounds) Browning bolt action rifle.**

The Pittman-Robertson Act

In 1987, Browning joined in the commemoration of the 50th Anniversary of the Pittman-Robertson Act, which is known officially as the Federal Aid in Wildlife Restoration Act. Pittman-Robertson may well be the single most important piece of legislation ever enacted to benefit wildlife.

Over half a century ago—in 1937—the Pittman-Robertson Act was sponsored by Senator Key Pittman of Nevada and Representative A Willis Robertson of Virginia. This act directs the proceeds of an 11 percent excise tax—on the manufacturing cost of guns and ammunition—specifically to programs involving wildlife management. The fundamental concept was that sportsmen would directly help pay for the needs of wildlife through monies collected by the tax.

This excise tax was heartily supported by sportsmen and the wildlife management community. Recent amendments have broadened it to include archery equipment and pistols as well, and each year, more than $100 million is collected from sportsmen through this tax. Over the past five decades of Pittman-Robertson, more than $1.5 billion has been collected for wildlife.

State wildlife agencies receive portions of these monies according to a formula, the criteria of which are the state's population, its general need and the number of hunting licenses sold in that state. The state proposes conservation or research programs and, if approved, Pittman-Robertson funds will cover up to 75 percent of the cost—or, in dollar terms, three dollars for every one dollar put up by the state.

More than 62 percent of the money collected goes directly into land acquisition. Since the program began, over 3.7 million acres have been purchased to develop and operate wildlife management areas. Purchasing land and improving habitat has always been of utmost importance, as habitat loss is the most significant cause of reduced wildlife populations. Twenty-six percent goes directly into research or surveys. A recent amendment allows about seven percent of the tax to be spent on hunter education programs. The remaining five percent goes to the states for planning and supervision, and for providing and distributing technical information to the pertinent agencies. Funds are used where they are needed most—for both game and protected animals.

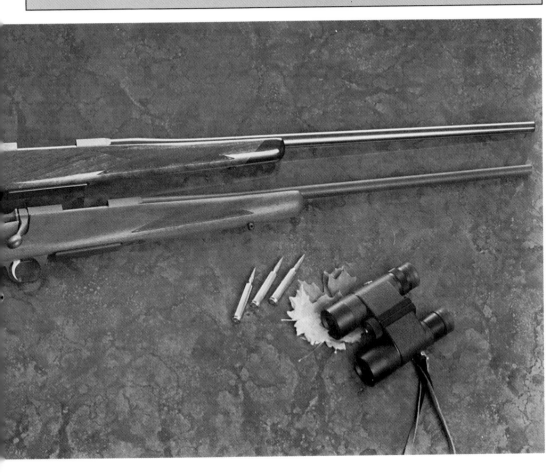

Newer Shotguns

Browning Shotguns

The following Browning shotgun was distributed by Browning of St Louis, Missouri, but was manufactured by Remington Arms Company of Ilion, New York. It is based on the original John M Browning autoloading shotgun design.

Browning Grade I Three or Five Shot Autoloading Shotgun. This is a recoil-operated autoloading shotgun in 20, 16 or 12 gauge, having a two- or four-shot tubular magazine, and a 26- to 32-inch standard barrel with any standard boring. This shotgun weighs from about 6.9 to eight pounds depending on gauge and barrel length, and has a checkered pistol-grip stock. It was available in Special, Special Skeet and Utility Field models. The Grade I Three or Five Shot Autoloading Shotgun was made from 1940 to 1949.

The following are major modern Browning shotguns which have been manufactured for Browning of Morgan, Utah, by Fabrique Nationale d'Armes de Guerre (now Fabrique Nationale Herstal) of Herstal, Belgium, and by Miroku Firearms Mfg Co of Tokyo, Japan. Many of these are either updates of or follow-ons to, original John M Browning designs.

Browning Automatic-5, Standard. This is a recoil-operated shotgun in Magnum 20, 16, Magnum 12 and 12 gauge, with pre-World War II 16-gauge guns having been chambered for 2.6-inch shells, and with standard 16 gauge guns having been discontinued in 1964. With a four-shot magazine and one in the chamber, prewar guns were also available in three-shot models. Barrels for these guns are from 26 to 32 inches in length, and barrel styles include plain, raised matted rib and ventilated rib with a choice of standard chokes. Weight, depending on stylistic treatment, runs from 7.3 to eight pounds. The Browning Automatic-5 can be found in Trap, Magnum 12, Light 12, Buck Special, Magnum 20 and Skeet models, and generally has a checkered pistol-grip stock. It was made from 1900 to 1973.

Browning Sweet 16 Automatic-5. This is a recoil-operated shotgun in 16 gauge, with a four-shot magazine and one in the chamber. Barrel lengths for this gun were from 26 to 32 inches, and included plain with striped matting on top of barrel, raised matted rib and ventilated rib styling. This gun weighed from 7.3 to eight pounds, with a lightweight version that has a gold-plated trigger. The Browning Sweet 16 Automatic-5 was made from 1937 to 1976.

Browning Superposed Hunting Shotguns. These are double-barrel, over and under, box lock shotguns, equipped with selective automatic ejectors and a selective single trigger (earlier models had double triggers, twin selective triggers or a non-selective single trigger). Bores run from .410 to 12 gauge, with any standard choke, and barrel lengths run from 26.5 to 32 inches. Styles include raised matted rib or ventilated rib, with the prewar Lightning model without barrel rib, and the postwar version of same supplied only with a ventilated rib, and weights run from 6.3 to 7.7 pounds. Stocks tend to be checkered, with pistol grip.

The higher-grade Superposed Hunting Shotguns—the Pigeon, Pointer, Diana, Midas and Grade VI models—differ from standard Grade I models in overall quality, engraving, wood and checkering; also, Midas and Grade VI guns are

Right, above and below: **Delicate scrollwork covers the polished grey receiver of this Grade VI Citori. The left side of the receiver features three mallard drakes. The Grade VI Citori can also be customized with a blued receiver and gold inlay.**

Browning and Bows

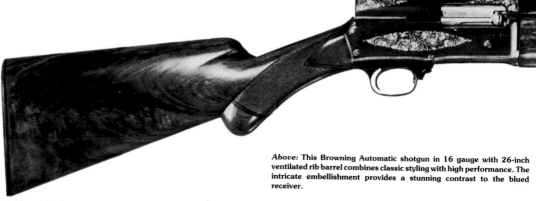

Above: **This Browning Automatic shotgun in 16 gauge with 26-inch ventilated rib barrel combines classic styling with high performance. The intricate embellishment provides a stunning contrast to the blued receiver.**

richly gold-inlaid. Browning Superposed Hunting Shotguns have been made from 1928 to the present.

Browning Double Automatic, Standard Grade (Steel Receiver). This is a short recoil system autoloading shotgun in 12 gauge only, having a two shot capacity, and barrel lengths from 26 to 30 inches. These guns are equipped with any standard choke and weigh about 7.8 pounds, having checkered pistol-grip stocks. They were made from 1955 to 1961.

Browning Twelvette Double Automatic. This is a short recoil system autoloading shotgun in 12 gauge, having a two-shot capacity, and having barrel lengths from 26 to 30 inches, with a wide selection of any standard choke.

This is essentially a lightweight version of the Double Automatic, with an aluminum receiver, and a weight of 6.8 to seven pounds, depending upon barrel length. Receivers for the Twelvette series were anodized in grey, brown and green with silver engraving. The Twelvette was made from 1955 to 1971.

Browning BT-99 Grade I Single Barrel Trap Gun. This is a single-shot, box lock shotgun with an automatic ejector mechanism, in 12 gauge, with a 23- or 24-inch ventilated rib barrel and modified, improved modified or full choke. It weighs about eight pounds and has a checkered pistol-grip stock with beavertail forearm and recoil pad. The BT-99 has been made from 1971 to date.

Browning BSS Hammerless Double Barrel Shotgun. This is a double barrel, side by side, box lock shotgun in 20 and 12 gauge, with automatic ejectors and a non-selective single trigger. Barrels range from 26 to 30 inches, with chokes ranging from improved cylinder and modified, to modified and full, to both barrels in full choke. Barrel styling is a matted solid rib, and weights are approximately seven to 7.3 pounds.

Stocks tend to be checkered pistol-grip, with beavertail forearm variety. This gun has been made from 1972 to date.

Browning Liège Over and Under Shotgun. This is a box lock superposed double barrel shotgun, with automatic

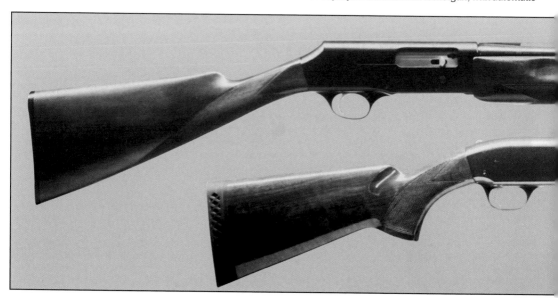

ejectors and a non-selective single trigger in 12 gauge. Barrel lengths range from 26.5 to 30 inches, with choke combinations of improved cylinder and modified, modified and full or both barrels full choke. The barrel style is ventilated rib-type, with gun weights, depending upon length of barrels, ranging from 7.3 to 7.9 pounds. A checkered pistol-grip stock completes the picture for this shotgun that was manufactured from 1973 to 1975.

Browning Citori Over and Under Hunting Shotgun. This is a double barrel, over and under, box lock shotgun in 20 and 12 gauge, having automatic ejectors and a selective single trigger. Barrels range from 26 to 30 inches, with choke combinations including improved cylinder and modified, modified and full and both barrels at full choke. Trap and Skeet models are major variants. With standard ventilated rib styling, weights, depending upon barrel length, run from 6.8 to 7.8 pounds. A checkered pistol-grip stock with recoil pad and semi-beavertail forearm are standard for this shotgun line that was made from 1973 to date.

Browning 2000 Gas Automatic Shotgun. This is a gas-operated autoloading shotgun in 20 and 12 gauge, having a four-shot magazine and barrel lengths from 26 to 30 inches with any standard choke. Plain matted barrel or ventilated rib are available, with gun weights ranging from 6.7 to 7.8 pounds, depending upon gauge and barrel. A checkered pistol-grip stock is standard, and this gun is also available in Field, Magnum, Buck Special, Trap and Skeet Models. The Browning 2000 has been made from 1974 to date.

Browning BPS Pump Shotgun. This is a pump action repeating shotgun in 12 gauge, with a magazine that holds five 2.8- or four 3-inch shells, and has available barrel lengths of from 26 to 30 inches. Chokes offered are improved cylinder, modified or full, with a ventilated rib barrel. This shotgun weighs 7.8 pounds more or less (depending on barrel length). It was introduced in 1977.

Browning Bows

Browning bows are designed by experts. Harry Drake, Browning's master bow designer, is well known to the world of archery. His Special Flight bow designs have dominated the National Archery Association Flight Championships since 1947.

On that date, the National and World Flight Record for the greatest distance an arrow was shot from a hand held bow was established with a Flight Bow designed by Harry Drake. The current World Flight Record for maximum distance from a hand held bow, in both men's and women's professional competition, was set with Drake flight bows.

In addition, every World Flight Record for foot bows since 1959 has been attained with a Drake design. Harry Drake also designed the compound hunting bow that shot a conventional 450 grain broadhead hunting arrow a spectacular 582.8 yards—the greatest flight distance ever recorded for a hunting arrow. This amazing feat was achieved in 1976, at the Utah State Bow Hunters' Annual Shoot.

An outstanding competitor in his own right, Harry Drake currently holds several National Archery Association flight

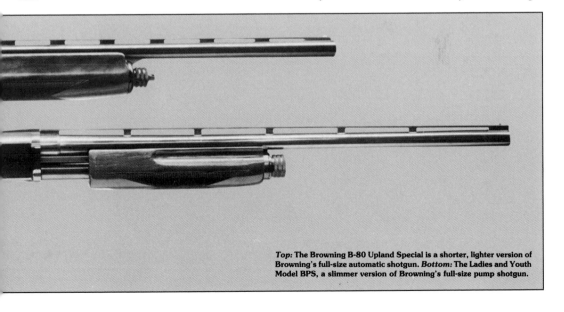

Top: The Browning B-80 Upland Special is a shorter, lighter version of Browning's full-size automatic shotgun. *Bottom:* The Ladies and Youth Model BPS, a slimmer version of Browning's full-size pump shotgun.

From top to bottom: The Limited Edition Superposed Pintail Duck Issue is hand engraved with delicate rosette patterns. Each firearm is inlaid and engraved with 18 karat gold. A handsome complement to this fine shotgun is the Limited Edition Black Duck Issue. Also shown are the Superposed Express Rifle, the Superposed Continental and rifle barrels for the Continental.

records, including the regular foot bow record of over 1542 yards. Harry also holds the honor of having shot an arrow further than any other man in the world. In 1971, using a Harry Drake bow in the Unlimited Foot Bow division, Harry shot an arrow 2028 yards, or 6084 feet! This earned him a place in the Guiness Book of World Records. More to the point, to design a bow and use it to shoot an arrow that far takes a thorough knowledge of how and why a bow limb works.

Browning bows are built for arrow speed, which depends entirely on how well the bow limbs can store and transmit energy. In theory, a bow's stored energy should be equal to, or greater than, the peak draw weight. In practice, this is rarely true. Friction in the cables and cams, combined with limb drag, can reduce a bow's energy by more than 10 percent.

Below: **In addition to its firearms, Browning also produces a line of bows, many of which are the inspirations of master bow designer Harry Drake. Known for their speed and accuracy, Browning bows are made for both hunting and range archery.**

To avoid this loss, Browning mounts its cams in limb notches rather than on bulky metal hangers. This helps the cams to operate smoothly, and doesn't add any unnecessary weight or drag to the limbs—thus, Browning bows utilize more of the bow limb's stored energy.

All Browning compound bows store more foot-pounds of energy than their respective peak draw weights. By comparison, when other bows of competing designs were tested, it was found that many of them stored less than their peak draw weights. On Browning 'four-wheelers,' the cables are attached to extending tuning bars, which allow the bow to be more accurately 'set,' or tuned, which in turn means more of the bow's energy is harnessed and not wasted.

Browning wood handle compound bows incorporate a warm, solid hand-filling riser that has been sculptured into a pistol grip. The African hardwoods that are used combine superior strength with the type of wood grain one is accustomed to seeing on high quality rifles and shotguns. These bows are easy to carry, weigh less than most other comparable bows and are easy to hold on target.

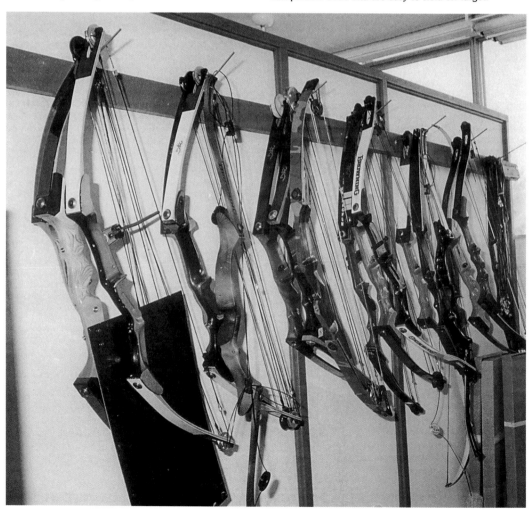

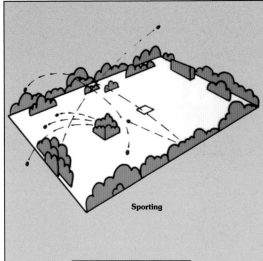

Sporting

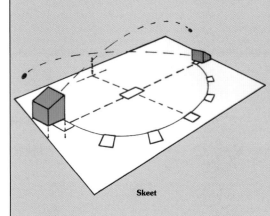

Skeet

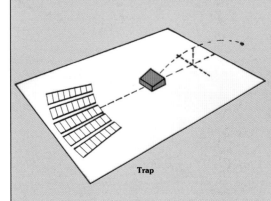

Trap

Shooting Sports

Non-game shooting sports involving shotguns fall into three main categories: Sporting, Skeet and Trap. The technical definition of each is as follows.

Sporting is of British origin, and has not yet been included in the roster of Olympic specialties, as trap and skeet shooting have. Nevertheless, this is a form of clay pigeon shooting that is rapidly gaining in popularity, as it closely replicates the true hunting environment. Clay traps are concealed diversely over a free course where they execute a plethora of throws, imitating high-flying pheasants; bolting rabbits; birds flying in front, and to the side, of the shooter; and game of various types skimming over the ground or rising up from brush. The action takes place both in the woods and out in the open.

Skeet shooting is practiced on a semicircular layout. A release hut is located at each end of the circle's diameter—a distance of 36.8 meters, or 40 yards. The left hut is termed the 'high tower,' and the right one the 'low tower.' The clay pigeons are shot at from eight stations in succession.

First the clay pigeon is thrown from the high tower, and then from the low tower, and so on. The shooter is entitled to one cartridge per clay, and each time, shoots from a different station on the circle. Shots from stations 1, 2, 6 and 7 involve 'doubles' in which clay pigeons are thrown simultaneously from the two towers. The shooter must try to break them with one cartridge per target, taking his first shot at the clay thrown from the nearest hut and his second at the one thrown from the farthest hut.

A complete series of eight stations therefore involves 25 clay pigeons. Skeet shooting is not only an official Olympic sport, but is also an excellent discipline for shooters who want to engage in their favorite sport outside of the regular game season.

Trap shooting is another Olympic specialty, but it is neither necessary to be a champion, nor is it necessary to have a costly trap range at one's disposal to enjoy trap shooting as a rewarding leisure activity.

Shooters are placed at a distance of 15 meters—or approximately 16.4 yards—behind a trench in which one or more throwing machines are installed. These machines are variously adjusted to provide a wide variety of clay pigeon trajectories. These trajectories are circumscribed by the limits established by International Regulations.

The starting order of the machines is purposely random, and the velocity, orientation and height of the clay's flight can vary with every shot. For Olympic events, 15 throwing machines are used; for Universal trench shooting, five machines are used; and for International ball-trap shooting, one machine is used. The shooter is entitled to 2 cartridges per clay. He changes his position for each clay pigeon. A complete trap shooting series is composed of 25 clay pigeons.

Craftsmanship

Craftsmanship and fine gunmaking are traditions at Browning. Take the exquisite work done on Browning Superposed shotguns for example. Each Browning Superposed is a unique masterpiece, in the best gunmaker's tradition, and a living proof of his love for his craft. Steels are selected for their high strength, taking into account the more than 24 different variants that are determined by the end-use of the part. French walnut is used for the stocks, having been selected for its density and the beauty of its veining and color.

The 84 parts that make up each Superposed are the result of 794 operations conducted on precision machinery; 64 of these parts undergo heat-treating according to their respective functions. In the course of the manufacturing processes, 1490 carefully-designed gauges and measuring tools are used to control 2310 dimensions. Machines then give way to the gunsmiths, who carry out by hand the striking off, fitting in, timing and finishing. The gunsmithing tradition is so strong in these craftsmen that, for example, each barrel filer makes his own tools—which he will then use for the external trueing of the barrels and for smoothing off the upper and side ribs.

This hand work confers individuality and excellence upon each gun. The gunsmiths who assemble the three principal elements of the gun—the action frame, the barrels and the foreend—use the technique of lampblacking, an ingenious method which is both simple and efficient. This ancient process is in fact the only way to 'see' the proper fitting and timing of the parts inside the closed gun. The gunsmith covers the parts in a fine layer of lampblack, or soot, produced by the flame of an oil lamp. He then fits them firmly together before taking them cleanly apart again. If there has been any friction between the parts, the lampblack will have disappeared. The gunsmith can then make any necessary modifications. This operation is repeated several times, until the fitting meets the very severe standards imposed by Browning.

In order to ensure the specified fit between the wood and steel section, the craftsman uses a technique similar to that involving lampblack, but this time, a special red dye is smeared on the action frame, which is then locked onto the stock. The gunsmith then disassembles the parts, and checks the places marked with red dye for proper fit, makes the necessary alterations, and thereby effects a precise fit.

One by one, with a watchmaker's meticulousness, each operating part of the firearm is carefully fitted. At the same time, stocks and foreends are rough cut, shaped, adjusted, carved and smoothed off. Each operation is carried out by hand, with a skill comparable to the fine materials being used. Only patience can succeed in the slow process of crafting steel and wood to a peak of artistry.

Then it's time for hand checkering; this is a process of regularity, precision and sureness of touch, and requires a true woodcrafter's knowingness. The surfaces that are to be engraved are then polished to an impeccable sheen, and the craftsman passes yet another work of art to those who will now adorn it with exquisite engravings.

Fabrique Nationale Herstal perpetuates Liège's traditional engraving of firearms. When asked by a visiting journalist to describe the most important personal quality of a good engraver, a Browning master engraver quickly answered, 'An iron constitution.' Engraving on steel does indeed demand, in addition to a very thorough artistic training, quite unusual powers almost 'spiritual' in their nature. It demands

Select walnut stocks and exquisite engraving exemplify the fine, traditional craftmanship found in today's Browning firearms. *At right, from top to bottom:* A Black Duck Issue Superposed, a Gold Classic Automatic-5 and a Pintail Duck Issue Superposed.

Craftsmanship Editions

concentration, calmness, sureness of hand and physical endurance to such degrees that only a few men are able to properly do the work.

It is an art that is executed with pointed chisels and a small, flat-headed hammer with which to strike them. With these, the craftsman carves out the drawing with a rare continuity and smoothness of line. It is a technique calling for absolute perfection, because one can not use an eraser on steel.

Browning offers an extremely wide array of classic, fine engraving styles, with accompanying backgrounds, figurative motifs, gold inlay and, in short, absolutely everything there is to offer in the way of firearms embellishment.

The fact that the over and under barrels give the Browning Superposed its characteristic elegance is the happy outcome of a number of functional advantages that result from its design, which has been widely copied. The arrangement of the barrels one above the other give the shooter a better view of the target, and allow faster and more precise aiming. No other double-barrelled shotgun will allow a well-aimed second cartridge to be fired so quickly after the first.

Special Editions

All major gun manufacturers produce special edition firearms and Browning is no exception. The firearms given in the following are among Browning's more famous special editions.

The Two Millionth Browning Automatic-5 Shotgun is a museum-quality firearm that was fit for a king and was actually produced for a President. In early 1970, the Two Millionth Auto-5 was taken from the production line and placed in the hands of Browning's very best Belgian craftsmen to be prepared for presentation to the then United States President, Richard M Nixon, who, under carefully-made arrangements, would then donate the gun to the Smithsonian Institution. Due to political considerations, the presentation to President Nixon and ultimately to the Smithsonian was never made.

The inlaid designs and patterns on the Two Millionth Auto-5 were created by Browning's master engraver, Andre Watrin, whose signature is engraved on the forward portion of the trigger guard. All gold inlay was executed by Master Engraver Jose Baerten and G Vandermissen. The designs on this unique firearm were inspired by an early exposition model Auto-5 that had been embellished by Browning's famous Master Engraver Felix Funken in 1930.

Technically, there is little engraving per se on this superb firearm, because all the decorative work was of 18 and 20 karat gold inlay. The left side of the receiver features a wreath framing the profile of John M Browning; above this is the description 'Browning Automatic Shotgun Number 2,000,000,' surrounded by elaborate border work and three mythological Chimera. The right side of the receiver features a wreath framing the symbol of the city of Liège, Belgium, in addition to elaborate scrollwork and eight

intricately inlaid mythological subjects of varying sizes. The top tang bears the gold inlaid signature of John M Browning, bordered by scrollwork and delicate gold borders.

The scrollwork featured on the receiver, tang, trigger guard and barrel is of 'fleur-de-lis' design. The gunstocks are of the highest grade walnut, featuring superb checkering performed by P Hanauer, Browning's best checkerer and woodcarver. The stock features a pistol-grip cap with a gold-plated oval inlay, bearing the inscription, 'Manufactured by Browning Arms Co, June 6, 1970. Invented by John M Browning, October 9, 1900.' Shipped to the United States on 26 June 1970, the gun was displayed in a special one-of-a-kind Browning Pro-Steel safe, featuring a handsome, full-color wildlife painting on the front door, a specially polished and finished interior, special lighting and a motorized rotating stand that provided a truly dramatic display.

It remained with Browning for some 15 years while the company sought an appropriate and meaningful use for this historic one-of-a-kind firearm. Then, in 1985, Browning offered the shotgun to the National Shooting Sports Foundation for display and auction at the Eighth Annual Shot Show in Houston, Texas. The proceeds from the sale went to the National Shooting Sports Foundation educational programs on safety, ethics and wildlife conservation in the shooting sports.

The Bicentennial Commemorative Superposed is a special limited edition issued to commemorate the United States Bicentennial. Fifty-one Guns were produced in 1976, each one of which celebrates one state of the Union and Washington, DC. The receivers feature side plates engraved with a gold-inlaid hunter and wild turkey on the right side, and the United States flag and bald eagle on the left side, with specific state markings inlaid in gold, and all on a deep blued background. The stocks and forearms are of American walnut, highly figured with decorative checkering. These are among the rarest of Browning firearms and some collectors value them today in excess of $10,000 each in new, unfired condition.

The Jonathan Browning Mountain Rifle commemorates Browning's roots in the Browning family. When Jonathan Browning invented his Cylinder Repeating Rifle, he designed a unique trigger system to operate it. During their development of the Jonathan Browning Mountain Rifle, Browning designers of the late 1970s found that, by making some modifications to this trigger system, they could give black powder enthusiasts a single trigger that functioned both as a standard trigger and as a highly sensitive 'set' trigger.

In addition to being remarkably fast and smooth, this newly patented trigger was amazingly uncomplicated. For the standard trigger, one had simply to cock the hammer

Opposite: **In contrast to the ornate engraving on the shotguns pictured on page 99, this superposed has a simple yet elegantly engraved receiver.** *Below:* **The Browning Auto-5 Ducks Unlimited 50th Anniversary Gun of the Year.**

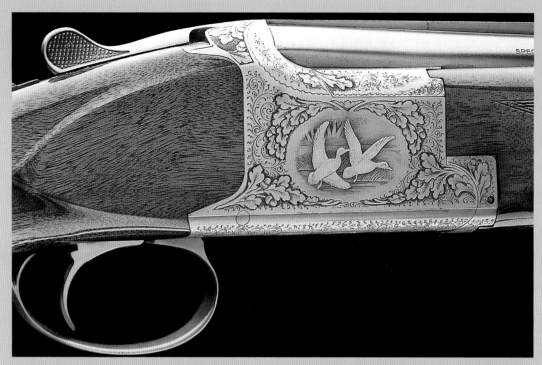

Above: This Presentation One Superposed Shotgun has oak leaf engraving on a silver grey receiver. Embellishing the sides of the receiver are two mallard ducks inlaid in gold. *Below:* This Presentation Two Superposed displays mourning doves in gold inlay on the receiver. The fleur-de-lis design on the receiver is more intricate than the engraving on the Presentation One.

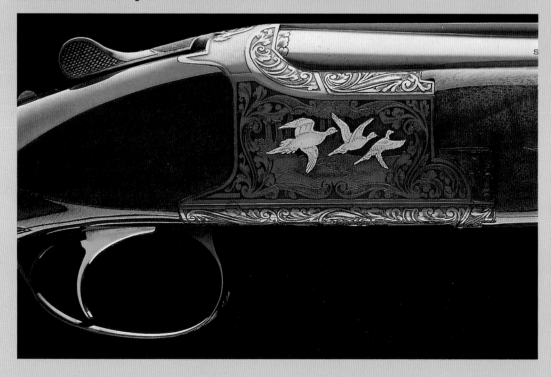

Above: The highly engraved receiver on this Presentation Three Superposed has a gold inlaid border and an English setter flushing a pheasant. *Below:* With its exquisite engraving and high quality finish, the Presentation Four Superposed is gunmaking at its finest. Deep floral carving adorns the receiver, which features a group of waterfowl in gold inlay.

At top, above and below: **These two guns illustrate the range of custom work. This BAR Big Game has a finely engraved silver grey receiver while the Grade I BAR has a simple, blued receiver.**

and squeeze the trigger. To use the 'set' trigger, the hammer was cocked and the trigger was pushed forward, thus sensitizing it to provide a light let-off that could be preset at between 2 ounces and 2 pounds.

The Mountain Rifle features traditionally-styled adjustable sights, which include a screw adjustment for elevation and drift adjustment for windage, with a sight radius of 21.9 inches, for accurate sighting. The Mountain Rifle is in .45, .50 and .54 caliber, weighs approximately 9.6 pounds and its hooked breech allows the barrel to be easily removed for cleaning. It is complete with a spare nipple and a ramrod of brass-tipped hickory.

Browning's unique, two-piece breech plug system reduces barrel stress and eliminates the danger of pre-ignition due to the presence of residue from previous firings.

The stock of a Jonathan Browning Mountain Rifle is specially selected and seasoned hardwood, finished with a deluxe oil finish and accented with a choice of browned steeled or brass finish on the butt plate, trigger guard and complimentary furniture. The barrel and lock are executed in traditional browned steel finish. A ram's horn on the

breech plug adds further distinction to this special edition black powder rifle.

The Limited Edition Waterfowl Superposed Shotguns were featured in Browning's 1980 catalogue. This special edition is devoted to celebrating the mallard duck, and features generous gold inlay life-studies of this regal American bird. Each shotgun in the series has a handsomely gold inlaid and engraved gray steel receiver.

Limited to 500 Belgium-made guns, the edition number of each firearm is written in gold on the bottom of the receiver along with the words 'American Mallard' and that bird's scientific name. Gracing each gun, in 24-karat gold relief against a grayed steel background scene, are the following depictions, which are based on drawings by western wildlife artists Leon Burrows. On the receiver's left side is a quiet pond on a fall morning with a pair of Mallards rising over it. On the receiver's right side, the birds come in over the calm waters of a marsh with wings set for landing. Another pair of Mallards rise to flight on the bottom of the receiver, and the golden head of a drake graces the bottom of the trigger guard.

The stocks of these distinctive firearms are beautifully carved, high grade dark French walnut with a hand-oiled

finish. The forearms feature extremely fine hand checkering, and a distinctive scroll pattern has been cut into the buttstock's rounded pistol grip. A uniquely checkered butt replaces the common or traditional plate. To compliment, display and protect these magnificent firearms, each commemorative Lightning Superposed was provided with a form-fitting, velvet-lined black walnut case.

The Classic and Gold Classic Editions, inaugurated in 1984, have as their focal points three of John M Browning's most appreciated inventions—the Automatic-5, the 9mm Hi-Power Pistol and the Superposed Shotgun.

The Classics are limited editions of 5000 units and feature engraved hunting and wildlife scenes on satin gray steel. Each model is individually inscribed with the engraver's name. In a banner across the right side of the receiver on the shotguns, and across the frame on the Hi-Power, are engraved the words, 'Browning Classic.' The left side similarly displays the words 'One of Five Thousand.'

Limited to runs of 500 pieces, the Gold Classics feature engraving similar to the Classics—with the addition of 18 karat gold inlays which include a portrait of John M Browning, the words 'Browning Gold Classic,' the issue number and each of the various wildlife that appear in the engraved

tableaux. These engravings and inlays are set against an exquisite background of grayed steel.

The Auto-5 was technically the first of the Classic and Gold Classic Editions. Its hand-selected, beautifully grained walnut stock and forearm feature a rounded semi-pistol grip that is similar to the traditional style found on early Auto-5s. The right side of the receiver on both Classic and Gold Classic Auto-5 models feature a pair of mallard drakes in flight, with the left displaying a labrador dog retrieving a downed duck.

The Classic 9mm Hi-Power Pistol was introduced in 1985. This ornate commemorative pistol features—on the top of its grayed, solid steel slide—an engraving of a bald eagle protecting her young from a lynx. On the pistol's right and left side are engraved profiles of an eagle's head, and the walnut grips have hand cut checkering and detailed, carved scrollwork. The pistol is complete with an attractive, velvet-lined display case.

The Classic and Gold Classic Over and Under Shotguns have hand-selected, highly figured, dense-grained walnut stocks featuring a traditional straight grip, English styling and Schnabel forearm with intricate hand-checkering. The grayed steel receivers on both editions have delicate scroll engravings and game scenes.

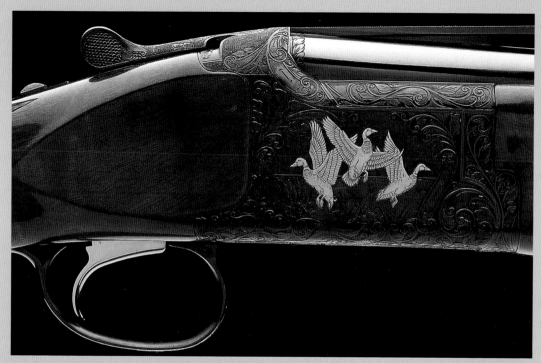

Above: The entire receiver, trigger guard and top tang of this Grade VI Citori feature deep relief engraving. Note that Citori's unique gold plating and engraving process captures an uncannily lifelike image of mallard ducks in flight. Below: The Gold Classic Over & Under with its highly figured receiver was made to commemorate the Superposed— John M Browning's last gun design.

The detailed engraving and gold inlay on the Black Duck Limited Edition Superposed *(above)* and the BAR Big Game Series Limited Edition *(below)* are fine examples of the high quality of firearms embellishment available from Browning's Custom Gun Shop in Belgium.

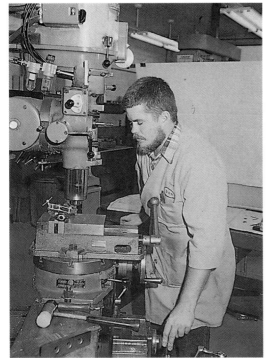

Above: The headquarters of Browning Arms Company in Morgan, Utah. The machine shop *(below left, below right, opposite)* produces some of the world's finest firearms. Here, the highly skilled workers carry on the tradition of gunmaking that began in the brilliant mind of the most prolific firearms inventor ever known, John Moses Browning. *Overleaf:* These Superposed Shotguns are a testimonial to the Browning tradition.

Index